The Exposome: A Primer

The Exposome: A Primer

the ex-POZE-ohm: a pr ĭ m′-er

the environmental equivalent of the genome

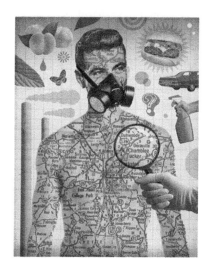

Gary W. Miller, Ph.D.
Department of Environmental Health
Rollins School of Public Health
Emory University

AMSTERDAM • BOSTON • HEIDELBERG • LONDON
NEW YORK • OXFORD • PARIS • SAN DIEGO
SAN FRANCISCO • SINGAPORE • SYDNEY • TOKYO
Academic Press is an imprint of Elsevier

Academic Press is an imprint of Elsevier
The Boulevard, Langford Lane, Kidlington, Oxford, OX5 1GB, UK
225 Wyman Street, Waltham, MA 02451, USA

First published 2014

British Library Cataloguing in Publication Data
A catalogue record for this book is available from the British Library

Library of Congress Cataloging-in-Publication Data
A catalog record for this book is available from the Library of Congress

ISBN: 978-0-12-417217-3

For information on all Academic Press publications
visit our website at store.elsevier.com

The Exposome by Michael Waraksa. First appeared in an article entitled, "Mapping the Exposome" in the September, 2013 issue of Atlanta Magazine. Copyright by Michael Waraksa.

This book has been manufactured using Print On Demand technology. Each copy is produced to order and is limited to black ink. The online version of this book will show color figures where appropriate.

Working together
to grow libraries in
developing countries

www.elsevier.com • www.bookaid.org

DEDICATION

This book is dedicated to my wife, Patti, and my six children for tolerating my near-manic writing sessions and providing me with a nurturing exposome.

CONTENTS

PREFACE

The idea for this book was conceived as the author organized and directed a course on the topic (Genome, Exposome, and Health) at Emory University in 2013. The course was based upon the research of faculty members at Emory University and published work from the primary literature. Students expressed an interest in having access to more background information about the course material. At the time no such material was available. A brief discussion with the editorial staff at Elsevier at a Society of Toxicology meeting led to a book proposal and a rapid turnaround of the text before you.

Dean Jones at Emory University, Eberhard Voit at Georgia Tech, and I have been involved in a series of collaborations over the past several years that touched on many of the same questions and challenges that surround the exposome. Dr. Jones and I were slowly emerging from our mechanistic biochemical and toxicological studies and embracing the computational, bioinformatic, and systems biology tools that were becoming necessary to unravel the scientific questions we were pursuing. Dr. Voit was especially helpful in opening my eyes to the importance of computational and systems biology. Many of these early discussions provided the impetus for the development of the course, as well as the proposal we developed for a center grant focused on the exposome. That proposal was ultimately funded by the US National Institute of Environmental Health Sciences in 2013. HERCULES: Health and Exposome Research Center at Emory provides conceptual and technical infrastructure for exposome-related research. A website, humanexposomeproject.com, has also been established to provide information to the lay and scientific public.

I would like to thank the scientists who participated in the inaugural course. Dr. Jones, Michael Zwick, Matthew Strickland, Jennifer Mulle, Lance Waller, Yang Liu, Yan Sun, Jeremy Sarnat, Dana Barr, Eberhard Voit (Georgia Tech), and Chirag Patel (Stanford). I am especially thankful to the students who participated in this inaugural course. They were very patient as we developed the course *de novo*. Their willingness to sign up for a course that included a word not even found in a dictionary or Wikipedia (at that time) is a testament to their inquisitiveness and openness to new ideas. An additional thanks goes out to Chandresh Ladva, a doctoral student in Environmental Health Sciences at Emory, for reading the manuscript and providing insightful feedback. I also want to thank the group of students in the Pharmacology and Toxicology Program at the University of Montana who viewed and critiqued the introductory lecture. Getting feedback from a group of students not associated with our institution helped with the further development of the course.

To faculty members attempting to integrate exposome-based concepts into their curriculum, I believe that you will find the topic to be one that engages and challenges the students (and yourselves). I hope that this introductory text makes it easier to do so. Even though it is in its primordial stage from a scientific perspective, the exposome has the potential to play a critical role in advancing our understanding of the environment in human health. Introducing the concept to the upcoming generation of scientists should instill a desire to better understand how the environment impacts health and hopefully inspire them to pursue the challenging questions surrounding the exposome. My hope is that you use the exposome to shamelessly and unabashedly promote the importance of the environment in health and disease.

I do not presume to be an authority on the exposome *per se*, but rather an environmental health scientist who is exceedingly interested in the concept. I undertook this project as a sole author because I thought it was more important to provide a focused and consistent, albeit idealistic, mindset throughout the book rather than provide an overwhelming, and potentially fractured, compilation on the topic. Given the early stage of the development of the exposome, it would be difficult, if not impossible, to generate such an authoritative tome at this time, even though such a work will be a welcome addition to the field. Certainly, such treatises about the exposome are forthcoming by those that are experts in particular aspects of the exposome, but at this stage it was my view that an introduction, or primer, was the most appropriate tack, that is, a course of action meant to minimize opposition to the attainment of a goal. This is necessary because there has been some reluctance, skepticism, and opposition to this topic. When one considers the potential utility of the exposome it becomes clear that this is, indeed, something that must be pursued. I have chosen to use the first person in those sections where I espouse my views and opinions that may not be consistent with others in the field, and I take responsibility for these thoughts. My goal is to engage trainees and colleagues so that they contemplate and critically analyze the exposome-related concepts and approaches.

CHAPTER 1

The Exposome: Purpose, Definition, and Scope

1.1 WHY A PRIMER?

Let me begin by explaining the title of this book. The exposome, which one can view as the environmental equivalent of the genome and an all-encompassing view of the exposures we encounter throughout our lives, is obviously the topic of the book, but why a primer? A primer is a small introductory book on a subject. The word, which has a short or soft "i," comes from the Medieval Latin word for first. It was not intended to be haughty or presumptuous, but rather exactly what it is—a small, introductory book. The connotation is distinct from that used when referring to paint (long "i"), but interestingly, that definition is also appropriate— the first coat. This book is for people who are interested in taking a first look at the exposome, to set the initial foundation for further study. The goal of this text is to provide an overview of the concept of the exposome and to explain how it can be used by students, scientists, physicians, and the general public to better understand the importance of the environment in health. Each chapter provides suggested readings for further study and discussion questions for further contemplation.

1.2 WHAT IS THE EXPOSOME?

The term exposome was first coined by Dr. Christopher Wild in a paper entitled, "Complementing the genome with an 'exposome': the outstanding challenge of environmental exposure measurement in molecular epidemiology." The paper was provocative and demanding. In it he put forth the first definition of the concept (Figure 1.1).

Original (Wild)

all exposures from conception onwards, including those from lifestyle, diet and the environment

Figure 1.1 Exposome definition as initially proposed by Dr. Christopher Wild. As noted in the text, Dr. Wild coined the term to represent the totality of our exposures.

Dr. Wild, an epidemiologist, expressed concern over what he perceived as a lack of information about our environmental exposures. Specifically, he was concerned with the lack of measures or tools that epidemiologists could use to help identify contributors and causes of human disease. Subsequent publications from Dr. Wild describe three distinct, but complementary, exposomes: the internal (the within the body measures), specific external (the immediate local environment, radiation, diet, lifestyle, pollution), and general external (societal, economic, psychological). Dr. Wild's idea was insightful and timely, but it has not received as much attention in the biomedical research community as one may have predicted and hoped. There are numerous explanations for this, but one possible reason was the initial focus as an issue for exposure assessment and epidemiology. Of course a measurable exposome would be immensely beneficial to these fields, but the potential impact of having a workable exposome goes far beyond exposure assessment and epidemiology.

Dr. Dean Jones, a colleague at Emory University and an expert in metabolomics, and I have proposed a slightly modified version of the aforementioned exposome definition. While many of these issues are addressed in Dr. Wild's papers, they are lacking in the protypical dictionary or textbook definition. Our suggested definition is as follows (Figure 1.2):

Refined (Miller and Jones)

The cumulative measure of environmental influences and associated biological responses throughout the lifespan including exposures from the environment, diet, behavior, and endogenous processes

Figure 1.2 An expanded definition of the exposome. Here, a new definition by Dr. Gary Miller and Dr. Dean Jones is provided that focuses on the measurable qualities of the exposome and emphasizes the importance of the body's response to the exposures that occur throughout one's lifetime. The exposome is a combination of the complex exposures and the complex responses to said exposures. The responses are arguably as important as the exposures themselves, as they represent the summation and integration of the exposures that occur within the context of our genetic mosaic.

The notable differences between the original definition and our refined definition are (1) *the cumulative measure*—that is the components that make up the exposome are measurable items and build up over time. If there are no current means of measuring, then it cannot be part of the entity. (2) *associated biological responses*-the way our bodies respond to various external forces are part of what makes up the exposome. The differential response seen among individuals may be as important as the exposure itself. For example, an environmental trigger may cause methylation and silencing of a gene, which dictates the ultimate outcome of that exposure. The alteration in DNA one measures at any given time results from the combination of damage from exogenous and endogenous sources, as well as the inherent DNA repair mechanisms. It is essential to include the body's response to the environmental insults. (3) Explicit insertion of the word *behavior*. Specifically, the word is used in a broad context here to capture behaviors that are self-initiated and those that are exerted upon us from outside of our bodies. This would include many of the social determinants of health, stressors that result from relationships and habitats, risky behaviors, and positive activities, such as exercise or physical activity. (4) *The endogenous processes*: This is important in that our bodies are an ongoing biochemical experiment. Thousands of biochemical reactions are occurring at any moment, from the breakdown of nutrients to the buildup of cells and tissues. These reactions yield many by-products that can impact our health. For example, such reactions generate free radicals that can damage macromolecules, but that also act as important signaling molecules. It should also be noted this expanded definition does not distinguish between internal and external exposomes or the concept of the eco-exposome (see Chapter 3). The exposome is by definition all-encompassing and includes elements internal and external to our body. Thus, while this expanded definition provides more detail, it is more inclusive and unifies several current schools of thought about what the exposome is (Figure 1.3).

Our refined definition sets the stage for a more comprehensive evaluation of the exposome and incorporates some key concepts that no longer need to be inferred from the original definition. We must be reminded that the exposome is massive and will resist any attempt to simplify its character. The following chapters will attempt to expound upon this initial definition by providing information about the various approaches and models that can be used to develop the structural framework of the exposome.

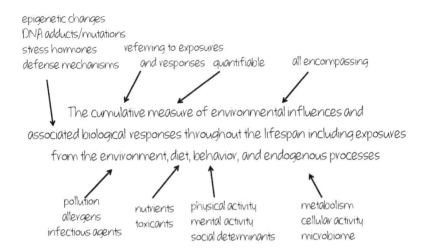

Figure 1.3 Anatomy of an updated exposome definition. The expanded definition proposed by Miller and Jones builds upon the foundation provided by Dr. Wild. Some of the key differences are shown here. The exposome is a measurable entity that will include a suite of measures. The environmental influences are all encompassing. The cumulative biological responses, that is, the results and responses to those exposures, are a key part of the exposome. Exposures to environmental factors, such as pollution, allergen, and infectious agents, as well as the beneficial nutrients and potential hazardous components of our diets, and the concept of behavior. By this we mean our activities (exercise, occupation, hobbies), our mental efforts (ongoing learning, engagement in intellectually stimulating activities), and social factors. The social factors are often not considered within the context of environmental health, but the stresses exerted by economic strains, community issues, and one's home and social life can have major effects on our health. Lastly is the inclusion of the endogenous processes. The processing and metabolism of chemicals necessary for function of cells and organs can generate harmful molecular species or be involved in various repair pathways.

1.3 DARWIN WOULD BE PROUD

Ever since humans (*Homo sapiens*) walked the planet, there has been an awareness of the influence of environmental factors to one's health. Even the Neanderthal (*Homo neanderthalensis*) could identify the predator as an external threat. Indeed, in those prehistoric times it may have been easier to recognize the exogenous forces acting upon the species than the endogenous factors. No textbooks were needed to convince the Neanderthal that getting hit in the head by a large rock was a bad thing. Nor was it necessary to be told to run when being pursued by a large reptilian. If a Neanderthal's associate consumed a plant and died, there is a good chance that our subject would avoid that poisonous plant in the future. Recognition of the adverse effects of external forces on our health has been intuitively obvious for millennia.

Move forward to more modern times. References to the manipulation of external factors abound, while the understanding of internal forces was somewhat limited until the past few centuries. While the practice of Mithridatism (more on this approach practiced by Mithridates VI of Pontos in the toxicology section of Chapter 4) likely induces some iatrogenic effects, these practitioners were concerned about an impending environmental insult and took action using a carefully prepared exogenous preparation. Something as simple as how blood flowed throughout the body was not understood until the seventeenth century when William Harvey proposed the model of circulation that revolutionized physiology and medicine. Prior to this, humors (black bile, yellow bile, phlegm, and blood), first introduced by Hippocrates and propagated through Galen, ruled the day. These four substances were thought to control our physiology. Even literature documents the importance of chemical manipulation of one's environment. From Chaucer's Canterbury Tales to Shakespeare's Romeo and Juliet, characters sought the counsel (and drugs) of their local apothecary ("though in this town there is no apothecary, I will teach you about herbs myself" to "O thru apothecary, thy drugs are quick") clearly recognizing that power of exogenous substances to influence health.

Centuries later the connection between exogenous germs and disease was revealed. Semmelweis's recognition of the need of surgeons to wash their hands in between autopsies and the delivery of babies, which was initially met with a great deal of skepticism, was ultimately confirmed by Pasteur and led to a revolution in the study of diseases of infectious origin, that is, exogenous. Insidious infectious agents are clearly a dangerous external force and should be included in the definition of the exposome. Up until late in the twentieth century, infectious disease and the environment were not commingled. It is more than the infectious agents and vectors that facilitate their spread being viewed as an environmental insult. Our environment and how we interact with it clearly influences the spread of disease vectors and our vulnerability to the pathogens. The emerging field of disease ecology is addressing this important relationship and will likely contribute to our understanding of the exposome.

From an investigatory perspective, genetics has a distinct advantage over the environment. The patterns of inheritance, gene replication, and regulation are wonders of nature and have been quite tractable (see Chapter 2). Although it is odd that the majority of the discussion of Darwin occurs among the geneticists when natural selection is primarily driven *by the environment*. The environment, or if the reader will allow, the exposome, is what is driving natural selection. Admittedly, the response of the organism and specifically its DNA is at the heart of the evolutionary process (i.e., the cumulative biological response), but Darwin's work is used in the classroom to lay the foundation for the field of genetics. What happened to the environment? Why is the Origin of Species not used to inspire students to study environmental health sciences? If one removes the external forces acting upon our genome, evolution will grind to a halt (Figure 1.4).

This brings us to the classical debate of nature versus nurture or genes versus environment. It is unfortunate that the word "versus" is even used when comparing the relative contributions of these two poles of the continuum. It should not be a competition. If one accepts it as a competition, one must acknowledge that the gene side has the upper hand and this may be why it is a larger driver of biomedical education. However, it is not an "either or" situation. Biology acts at the interface. There is a dynamic interaction between our genes and environment with a complex level of interdigitation (more detail in Chapter 2). Admittedly, there is somewhat of a competition when one looks at resources at national and institutional levels. Thus, if we are

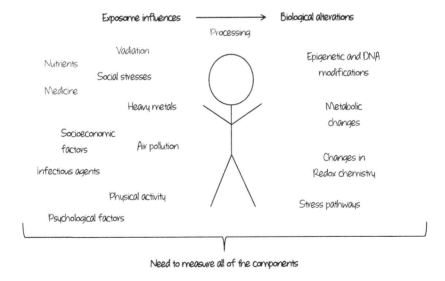

Figure 1.4 Multiple influences versus multiple changes. The exposome includes a vast array of external forces acting upon our bodies. The exposome requires knowledge of those external forces and the responses of the body to those forces.

going to view this as a competition, the environmental side needs a stronger proponent. We must be better prepared to properly represent the environment at a conceptual level. As noted in Chapter 2, a purely genetic approach to studying human disease is limited. The environment must be incorporated into our conceptual models of disease and health. The nature side of the discussion is very well represented. I would like to suggest that the exposome provides a quantifiable aspect of nurture. When viewed as a critical component of nurture, the exposome assumes the role of conceptual advocate for the environment.

1.4 IF IT IS SO OBVIOUS, WHY THIS BOOK?

If the association between external factors and human disease is so obvious, why do we need the exposome? Over the past century, modern medicine has produced some extraordinary advances. Surgical excision of tumors, repair of broken bones, development of antibiotics, noninvasive imaging techniques, robotic surgery, and laser-based eradication of tumors are, for the most part, innovations focused on fixing. Most modern health care systems are built upon this repair-based model, and research and development will deliver goods that fit into the model. While the dramatic external influences (car accident, gun shots) are readily apparent, the subtleties of environmental influences have slowly slipped in recognition. Environmental health is not taught in most medical schools. Toxicology instruction is confined to the understanding of the adverse effects of the medicines prescribed by the doctors, which is disconcertingly ironic. Doctors treat sick patients. They are certainly concerned with the obvious external forces that cause illness or injury (infection, trauma), but the less obvious external forces are underappreciated. Cumulative low-level exposures to chemicals in the workplace or home are simply not on the average physician's diagnostic radar. We know that many cancers are caused by exposure to toxicants in the workplace, but how often do you hear a physician inquire about one's environmental exposures (smoking is an exception, but consider how acceptable smoking was within the medical community just in the past century)? It has been estimated that if the practice of cigarette smoking (definitely part of the exposome) was eliminated throughout the world, it would have a more positive benefit to humankind than all of the advances of modern medicine combined (save vaccines). Control of the external forces represents a very powerful lever for improving health.

I am sympathetic to educational institutions that train our physicians and allied health professionals because it is exceedingly difficult to teach this material. Unfortunately, there is not a paradigm or framework that allows this information to be easily integrated into the current medical school and allied health sciences curricula until now. Enter the exposome. Perhaps I am overly optimistic, but I believe that the general concept of the exposome provides the framework to educate medical and health professionals and to inform the general public. The exposome is not at all complicated as a general concept. It is actually quite simple—the totality of exposures and how we respond to them. We must consider them all.

As simple as the general concept of the exposome is, the measurement and interpretation of the exposome is quite a different issue, and this is where the more specific definition in Figure 1.2 comes into play. Conceptually, it is appropriate to refer to the totality of our exposures, but when it comes down to measuring the exposome we should use a specific definition. The level of complexity is extraordinary. As detailed in Chapter 5, the mathematical approaches to the exposome will exceed those used by the field of genetics. Even with the perplexing chasm that exists between the simple concept and the complex definition, there is attainable middle ground. Any movement from the simple concept to the complex definition is positive. Bringing more environmental health into human health provides an exciting opportunity to improve how we address health care.

Many argue that we do study environmental influences of health and point to the efforts of the US National Institute of Environmental Health Sciences (NIEHS) or the Medical Research Council (MRC) Toxicology Unit in the UK. Let us look at the organization of the National Institutes of Health (NIH)—20 different institutes and 8 different centers that study disease ranging from autism, cancer, diabetes, cardiovascular disease, and Alzheimer's disease. NIEHS represents only 2.2% of the entire NIH budget (2.47% if one includes the Superfund allocation) and is located more than 250 miles away from the Bethesda campus, the Mecca of biomedical research in the US and the region of the country where all other branches of the NIH reside. While its ultimate proximity to the Environmental Protection Agency and other top research institutes in Research Triangle Park has some advantages, one cannot help but view its initial placement as a symbol of subtle ostracism. To take an even more cynical view, one could view this as a tacit bias against the importance of the environment at the national level. NIEHS should have its tentacles in nearly all of the other branches of NIH. It should have an advisory role across all of those institutes that would benefit from an enhanced understanding of the environment (I would argue that is all of them). Of course the other 19 institutes are free to allocate funding to study environmental causes of their diseases of focus, but this rarely occurs (the percentage of funding in environmental epidemiology, exposure science, and toxicology at institutes outside of NIEHS is meager). For the most part, the various institutes shy away from tackling challenging environmental problems expecting NIEHS to deal with them. Several years ago a Gene–Environment Initiative was launched at NIH; this was a welcomed collaboration that involved several institutes and laid the groundwork for the eventual Tox21 project (see Chapter 4), but it was still rather limited in scope. The priority of funding from NIH and other worldwide funding bodies (European Union, MRC) is important in that it drives and directs careers of scientists. Robust funding attracts top minds.

1.5 ENVIRONMENTAL HEALTH PRACTITIONERS

Where do those studiers of the environment roam? One can stratify these scientists based on their work setting and their discipline. Investigators interested in the environment generally work in academia, government, industry, or nonprofit agencies. Those in academia rely on the government for grant support and trainees for intellectual input and productivity. Governments support research and regulatory activities, which can have a major impact on public health. Industry scientists are often focused on minimizing the environmental impact of their particular sector or products. Those working for nonprofits are often focused on implementing practices that improve the environment in ways that enhance human health. For the most part, scientists interested in environmental factors in disease and health fall into one of three categories, which are covered in more detail in Chapter 4. Those that study these relationships at the population level tend to be epidemiologists that specialize in the environment. Those that measure and evaluate specific environmental exposures are exposure assessors or exposure scientists. Those interested in how the environmental contaminants impact specific biological molecules or pathways tend to fall in the domain of toxicology. These three distinct disciplines use different tools and approaches. While there is some overlap, the environmental epidemiologists are looking for interactions at a population level, the exposure scientists at a subset of the population or at the individual level, and the toxicologist at the molecular/cellular level (much work is done on a whole animal level, but rarely on an individual human).

In general, there is not a great deal of interaction among the three entities, but if progress is going to be made on the exposome this must change. These three core disciplines, which exist among a wide variety of employers, will likely drive the field with the input of specialists in systems biology, bioinformatics, genetics (yes, genetics), chemists, behavioral sciences, and many other disciplines. Thus, it is critical to get investigators outside the three core subdisciplines to view the exposome and its associated problems as an attractive scientific challenge. As noted in Chapter 5, the computational and bioinformatic approaches needed for the exposome are exceeding difficult, but provide a challenging and potentially rewarding pursuit for an enterprising and mathematically inclined investigator. It is incumbent upon those in the core subdisciplines to entice those with the necessary skills to collaborate on these projects or progress will be exceedingly slow.

1.6 THE EXPOSOME AS AN EDUCATIONAL TOOL

As noted above with regard to physician education, the exposome can also be viewed as a paradigm for teaching the importance of the environment in our health and Chapter 6 will explore this in more detail. A single undergraduate course could help the student place the myriad of exposures into the proper context and introduce them to historical and biological concepts and approaches. The concept of the exposome demands that the individual consider all of the forces that are impacting their health, not just the obvious issues that were made apparent by trips to the pediatrician. While genetics and socioeconomic backgrounds are nearly impossible to alter, as a young adult matures their activities, careers, behaviors, diet, and habits are under a considerable degree of control. This is a time where a student is developing independence and deciding their future pursuits. Framing their career and lifestyle choices within the context of the exposome provides an excellent foundation. For the graduate student, the exposome helps place their research into the bigger picture of environmental health sciences. It stresses the need to communicate with scientists outside their immediate discipline, and the ever-increasing importance of big science within the scientific enterprise. Encouraging this type of collaborative research should occur early in the formative years of the science trainee. The exposure scientist must gain an appreciation for the biological mechanisms of disease and the epidemiological methods employed in population-based research. The toxicologist must understand how basic research is applied to the human condition. The environmental epidemiologist must assure that the associations are grounded in biological pathways. Each must continue to expand their disciplines that may lend new tools and approaches for exposome-related research. For the medical student, it provides a comprehensive model for consideration of environmental impacts on disease and health. When diagnosing a chronic condition, the physician must understand how environmental factors impact the patient and be able to explain the multitude of forces to the patient. The exposome may be able to provide this framework to enhance the doctor–patient relationship. For the established environmental health scientist, the exposome represents the future—the inevitable need to deal with the massive data sets that result from ever-advancing technology. While this flies in the face of the reductionistic approaches that are taught in graduate schools, it is imperative that we counterbalance our reductionism with thoughtful construction of complementary theories and hypotheses. Most scientists are reductionists by nature and produce the bricks, but somebody has to assemble them (see Chapter 5). The exposome mandates that the process of building the structure is equal to that of generating the bricks. For the general public and community, the exposome provides a practical mental model for one's health as noted for the undergraduate student—an emphasis on the whole rather than the parts.

Regardless of one's place in the educational process, the exposome has value as a concept. The present work delves deeper into the practical issues that must be addressed to make the exposome valuable as a scientific construct. I encourage you to read the following chapters and suggested readings, and supplement these activities by searching the current scientific and lay publications for information that can help inform our study of the exposome.

1.7 DISCUSSION QUESTIONS

How do the students perceive the balance between nature and nurture? Which is more important? What aspects of nurture does the exposome miss? Does this vary among different conditions and diseases? What are the scientific methods or approaches to measure the two sides of the nature:nurture equation?

FURTHER READING

Miller GW, Jones DP. The nature of nurture: refining the definition of the exposome. Toxicological Sciences 2014;137(1):1–2.

The National Academies. The Exposome: A Powerful Approach for Evaluating Environmental Exposures and Their Influences on Human Disease. Washington, DC: The National Academies; 2010.

Wild CP. Complementing the genome with an "exposome": the outstanding challenge of environmental exposure measurement in molecular epidemiology. Cancer Epidemiol. Biomarkers Prev. 2005;14(8):1847–50.

CHAPTER 2

When the Genome Falls Short: Limitations of a Gene-Centric View of Health

2.1 DNA, NO LONGER A SECRET

The discovery of the secret of life as announced by James Watson and Francis Crick at the Eagle Pub in Cambridge ranks high in the annals of scientific history. Their elucidation of the structure of DNA stands out as one of the greatest scientific discoveries of all time. As stated in the opening lines of the 1953 *Nature* article "we wish to suggest a structure for the salt of deoxyribose nucleic acid (D.N.A.). This structure has novel features which are of considerable biological importance." This surprisingly reserved statement from the pair, not known for being subtle, foreshadowed a revolution in biology. It is difficult for scientists in the twenty-first century to fathom life without knowing the structure of DNA. It is taught in grade schools, but just over 60 years ago the term "double helix" was not in the scientific vernacular. While DNA was known to be the genetic material there was no understanding of how it transferred the information. The penultimate line penned by Crick was similarly prophetic, "It has not escaped our notice that the specific pairing we have postulated immediately suggests a possible copying mechanism for the genetic material." In follow-up papers Watson and Crick describe the pairing mechanism that allowed for the DNA to serve as a template for gene copying. These findings represented a paradigmatic shift in the truest sense of the term later made famous by Kuhn. The recent auctioning of Francis Crick's Nobel Prize (USD 2M) and letter to his son describing their discovery (USD 6M) bears witness to how extraordinary this discovery is viewed in modern scientific history. Of course there were many other scientists that contributed to the knowledge that permitted the discovery, from Wilkins and Franklin, to Bragg and Pauling, it was a stunning discovery based on a combination of thinking and tinkering that would forever change biology. The impact of this discovery and the implications of it would transform biomedical research.

Move forward 50 years to the completion of the first draft of the Human Genome Project in 2003. This was no two-man show tinkering with models. The Human Genome Project was big science; billions of dollars, hundreds of scientists, and 3 billion base pairs. While the 1986 Sante Fe conference is often considered the kickoff to the human genome sequencing project, one might say it started back in Cambridge. The Human Genome Project required some major technological advances. Indeed without the development of sequencing by Gilbert and Sanger (for which Sanger won his second Nobel Prize. Really? Two?) and its further automation by Leroy Hood there would be no initiative. This technological advance spurred scientific water cooler discussions of sequencing the entire human genome. The often underappreciated efforts by the US Department of Energy with their interest in studying the mutagenic effects of radiation played an important role in generating key infrastructure and knowledge. Technology was key. Another example of a critical technological advance was the discovery of the polymerase chain reaction (PCR). Most scientists take the routine technique of PCR for granted. Scientists had been working with DNA for decades, but the ability to produce large quantities of specific sequences was technically challenging. Enter *Thermophillus aquaticus* and a quirky, some say bizarre, surfer/scientist from California. Dr. Kary Mullis was a scientist at Cetus Corporation where a group of investigators were working on methods of DNA copying. But it was Mullis who, not unlike Watson and Crick, assembled the *ex post facto* apparent pieces to solve the riddle (and be awarded a Nobel Prize). Mullis attributes some of his creative thinking to his prior (heavy) use of lysergic acid diethylamide (LSD), which may also have contributed to some of his subsequent pseudoscientific ideas. However, it is also highly likely that input from other Cetus scientists was critical in the formation of his breakthrough discovery.

Being able to manipulate large pieces of DNA was enhanced by the development of bacterial- and yeast-based artificial chromosomes (BACs and YACs). YACs were first described by Andrew Murray and Jack Szostak (Szostak went on to win the Nobel Prize in Physiology and Medicine for the codiscovery of telomeres, more on this later). These systems allowed the genome to be fragmented into more manageable parts and be grown in cell-based systems. YACs are composed of an artificial chromosome that contains a centromere, telomeres, and a replication origin element, which allows replication in yeast cells.

YACs can be used to clone fragments from 100 to 3,000 kb. YACs do have a tendency to be unstable and can introduce errors. This was why the Human Genome Project transitioned to the use of BACs for the remainder of the project. BACs have a similar composition to YACs and have become a mainstay in the study of genetics. First reported by Melvin Simon and Hiroaki Shizuya at the California Institute of Technology, BACs are excellent vector systems for the chromosomal DNA libraries. Besides their utility in sequencing projects many scientists are also aware of their utility in the production of transgenic mice. Being able to include much larger sections of DNA than previously possible allows large sections of genomic DNA to be transferred to the donor embryonic stem cell, including the ability to employ endogenous promoters to drive expression. This is especially helpful when modeling complex genetic disorders that involve regulatory units that span large distances on the chromosome.

Armed with the aforementioned significant technological developments and support from international heavyweights, such as the Wellcome Trust in the UK and the US NIH and Department of Energy, the field moved forward with this Herculean project. Major centers involved in the project include the Whitehead Institute, the Sanger Centre, Washington University in St. Louis, and the Baylor College of Medicine. The NIH-side of the effort was headed up by none other than James Watson. Watson provided a high level of enthusiasm and gravitas to the project. Over the course of the next several years, excellent progress was made. The government-funded initiative was proceeding slightly ahead of schedule. Then enter one Craig Venter from The Institute for Genome Research (TIGR) and his newly formed company Celera Genomics (celera means swiftness in Latin). The brash scientist, who had previously worked on the Human Genome Project while employed at NIH, proposed to use a shotgun approach to complete the sequence of DNA in a mere 3 years (4 years faster than what the Human Genome Project had planned). Who says science is not fun? His new spinoff company, Celera Genomics, had raised a substantial amount of private funding and proposed to do the job better and faster. As a budding scientist I recall the race for the sequence of DNA and was entertained by the sheer audacity of Venter. While there was concern that Venter's approach was scientifically more coarse and perhaps professionally more crass, it was clear that a sense of urgency had developed among the Human Genome

Project teams. This is a wonderful illustration of the value and travails of competition in science. Venter and his company eventually partnered with the Human Genome Project team to complete the project together, but the acceleration of the project would not have occurred without the tacit and overt sense of competition from Venter's organization. It was great scientific theater.

One of the issues surrounding the efforts of Celera Genomics was the intent to patent genes and genes sequences. This was something Watson opposed with a passion. Indeed, Watson stated that patenting DNA was lunacy.[1] Ultimately, this disagreement led to Watson's resignation in 1992. Francis Collins soon took over as Director of the Human Genome Project. After a series of bigger-than-life DNA personalities, it has been comforting to see a seemingly even-keeled and humble Francis Collins get the Human Genome Project to the finish line and become the leader of the NIH. Dr. Collins has compiled a list of important lessons learned from the Human Genome Project. In a paper entitled, "The Human Genome Project: Lessons Learned from Large-Scale Biology," Collins made it clear that big science is different from the traditional way of doing science. Indeed, this paper, which we will revisit in Chapter 7, is a gift to the exposome initiative. Many of these lessons are directly applicable to steps that would need to be followed to pursue a human exposome project.

Genome-wide association studies (GWAS) were hailed as a novel way to identify genetic associations with various disorders. Data are displayed in the familiar Manhattan plot design where the landscape shows associations by taller peaks at given chromosomal locations. In seconds, the human mind can scan the association with the aid of the colorplot. There have been numerous successes with novel associations and pathways identified, but the relatively low resolution based on single nucleotide polymorphisms (SNPs) has limited the utility of the GWAS. Many studies have been criticized for study design and lack of appropriate clinical characterization of the patients, but these concerns hold true for any type of research conducted on a human population. A 2012 opinion piece in the *Journal of the American Medical Association* lays out the promises and limitations of GWAS, but

---

[1]In 2013, the US Supreme Court agreed with Dr. Watson by nullifying Myriad Genetics' patent on the gene associated with breast and ovarian cancer, *BRCA1*, but left the door open to the patenting of DNA sequences not found in nature.

foreshadows its demise. As another example of the swiftness of science, the authors suggest that exome or deep sequencing, where the complete sequence of a suspected chromosomal region is obtained, will soon supplant the GWAS (although some of the same statistical approaches will be used). Full genome sequencing will likely be available and affordable shortly after the publication of this book. Availability of each person's genome could be viewed as the true completion of the Human Genome Project. Yet, even armed with the complete genome, major impediments to understanding disease persist. Three billion base pairs—that is a big number. The computational power required to conduct such analyses continues to grow and strategies to tame these types of data sets are discussed in Chapter 5.

Another key genetic development was the International Haplotype Map Project (HAPMAP). In this initiative, started in 2002, the research team set out to assess the variability in the human genome. Since the Human Genome Project involved the sequencing of a small pool of human DNA, the initial sequencing could not address the diversity among the human population, let alone how those variations contributed to human disease. Therefore, the HAPMAP project aimed to assess how much the human genome varied across people throughout the world by sampling a larger and highly diverse population for analysis (from an original ancestor approximately 100,000 years ago). SNPs represent areas of the DNA that are variable. An SNP chip may have 1 million different SNPs, but this represents a mere fraction of the genome. By obtaining data on the relative incidence of SNPs among different human populations it has been possible to identify points of divergence and various disease associations. This even includes the determination of one's percentage of Neanderthal DNA (the author's is way above average—94th percentile. Thanks, 23andME—see below).

One of the concerns with GWAS and SNP-based studies is that SNPs shown to be associated with a certain condition may be misleading. Linkage disequilibrium is where two SNPs have been inherited together because of their chromosomal proximity. Thus, SNP 123456 may be responsible for an increased incidence of disease X, but not be on the SNP chip (because only a small portion of the genome is represented). It is possible that SNP 123000 and SNP 124000 have been inherited together with 123456 over time and, thus, show up positive for the association with disease X, but only because they are located near the culprit SNP 123456. Not having SNP 123456 on the chip and only having information from the flanking SNPs would yield a misleading low intensity effect with a low odds ratio (OR). But if the experiment was repeated with a chip that contained 123456 the OR would rise significantly. This is an argument for deep (or complete) sequencing, where the entire portion of that particular gene is sequenced. The main point is that SNP-based approaches do not provide a high enough degree of resolution to pin down some specific associations.

As of this writing, the company 23andMe will measure over 1 million SNPs for only 99 USD. Risk alleles for disease conditions can be determined by submitting a sample of saliva. 23andMe uses the same type of SNP chip that has been used in GWAS studies. For example, if a study shows that people with a particular allele have a 20% increased risk of disease, 23andME would assign a 20% increased risk *based on that particular allele*. 23andMe is unable to assess other factors that impact that disease. Statistically, 23andMe is correct in assigning this 20% increased risk, but this is accurate at a population level. A person with this inherited 20% increased risk may have a handful of positive risk factors that actually make that individually 30% less likely to get the disease. The provided values are only measuring genetic risk and the company repeatedly reminds the consumer of this, but the participant does not have access to an equivalent value of their environment or risk factor impact. Individuals can be provided with genetic cardiovascular risk factors, but if the consumer does not have access to information on what actions can influence these risks, the impact on health is minimal. Ideally, the consumer would be provided with an analysis of the modifiable risk factors and how they can counterbalance the genetic risks, but at this time this type of analysis is not available.

2.2 THE GENE VERSUS ENVIRONMENT CONTINUUM

It is relatively easy to evaluate the causes of disease at the two ends of the continuum. Autosomal dominant or autosomal recessive diseases give rise to conditions that have been followed through family pedigrees for centuries. Some of these disorders have partial penetrance and may not show up in every generation, but the genetic link is still obvious and environment appears to play little role. On the other end of the spectrum, conditions due to completely external forces, such as head trauma, asphyxiation, or even shark attacks appear to have no genetic component. The role of one's genetic background in response to a car accident is minimal, although the genetically endowed reaction time, visual acuity, and physical and mental dexterity may play a role in the cause of the incident and clotting profiles may impact the response and survival, but for the most part the incident is environmental (Figure 2.1).

Back on the genetic side of the continuum, there are some instances where the environment does play a role. The age of onset of disease can vary among these genetic disorders, perhaps through environmental influences. An interesting example is phenylketonuria (PKU). PKU results from the mutation of the gene that metabolizes phenylalanine, one of the 20 amino acids that serve as building blocks for our proteins. Until 60 years ago PKU was a leading cause of mental retardation. Once it was discovered that the disease was the result of altered amino acid metabolism, the medical field was able to develop a test and treatment. While it is not trivial to follow, the treatment is somewhat simple: severely limit the intake of phenylalanine in the patient's diet. Or to put it another way, *modify the patient's environment.* This insanely strict diet followed with precision entirely prevents the adverse effects on brain function. A mutation in a single gene causes a

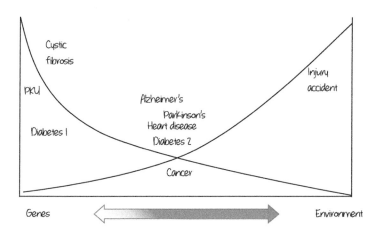

Figure 2.1 The gene–environmental continuum. There are numerous diseases that result exclusively from genetic abnormalities. These are depicted on the left side of the graph. There are also outcomes that result exclusively from external or environmental sources, shown on the right side of the graph. The vast majority of disease though resides at the interface. They may be 80% genetic and 20% environmental or vice versa. The past few decades have generated superb data on the genetic causation of disease. In order to address the majority of diseases at the interface we must have more comprehensive environmental data (i.e., exposome).

very serious disease that can be completely treated by altering the environment. While this stands out as one of the clearest examples of the ability of the environment to trump the genome, the formula may hold true for many other disorders.

The picture gets cloudier when we move away from the extremes. Complex disorders like heart disease, diabetes, and Alzheimer's disease have multiple genetic and environmental contributors. There could be a dozen different genes that confer a particular level of risk in an individual and dozens of environmental factors that interact with those genetic factors. As noted above, the field of genetics is making great progress on identifying the multiple genetic factors that may contribute to these diseases, but without a similarly detailed analysis of the environmental contributors the puzzle will remain unsolved. The study of complex human diseases must examine genetic and environmental influences and do so in a systematic fashion.

2.3 A DANGEROUS METAPHOR?

The balance between genes and environment has often been summarized by a common phrase bandied about in environmental health sciences. It is not totally clear who first stated it, but the author has heard it from two directors of the NIEHS, as well as Dr. Francis Collins, Director of the US NIH. The first reference to it occurred in the 1920s by Elliot Joslin (Figure 2.2).

"Genetics loads the gun, environment pulls the trigger"

Figure 2.2 The gun analogy. This analogy has been used for many years to describe the interaction between genetic and environmental factors. Perhaps because of the difficulty in modifying our genetics, the environment has been viewed from the same static perspective.

This author has never been fond of this phrase in that is has a very fatalistic tone. It is as if the only thing the environment can do is pull the trigger. It suggests that all environmental influences lead to catastrophic damage. This is not the case. Our environment can have beneficial effects on our health and can even contribute to repair after damage. Indeed, Dr. Joslin was a key proponent of environmental modification as medicine and the Joslin Diabetes Center at Harvard University has been a leader in the use of environmental modification to treat the disease. More importantly, however, is that careful regulation and manipulation of our environment can tie a knot in the barrel, rendering the genetically endowed predisposition essentially moot. Our destiny is not a loaded gun. Adoption of practices that minimize our exposure to adverse factors and maximize our exposures to positive factors can have major impacts on our health. The more appropriate phrase is given in Figure 2.3.

"Genetics loads the gun and the environment may or may not increase the likelihood of the trigger being pulled, but the environment could also influence whether or not the bullet remains in the chamber or in the event that the trigger is actually pulled the environment could still influence the path of the projectile and even be involved in the response to the impact."

Figure 2.3 The gun analogy. An expanded and intentionally absurd version of the analogy. The environment is not a unidirectional influence. The environment can do more than push the rock down the hill.

Alternatively, we could just stop using the original phrase and focus on the modification of our environment in a way that makes us healthier.

Let us move forward, say 20 years, when it is possible and affordable to conduct a study on a population of 1,000 with complete genome sequencing. We know all 3 billion base pairs for each of the 1,000 participants. Based on our knowledge of genetics, we will be able to predict which people are more likely to develop diabetes, cancer, or heart disease based on their genetic risk. But what then? Let us refer back to Darwin's natural selection. Natural selection is a response of the species (and the DNA) to external forces. It is the external forces that drive the variation. Thus, in our study population of 1,000, we can make all sorts of predictions of who will get what. The genetic data are fairly precise. But we know that all of the predictions are statistically based and assume a normal distribution. Yes, on average a certain allele will increase the incidence of x within a certain population, but it is not *ipso facto* for each individual. Manipulation of one's environment can alter one's presumed DNA destiny. The predictions will fail with a fairly high rate. Why? Because of the individual variability exerted by the external factors that make up the exposome. External forces acting upon the individuals will alter their genetically predetermined path. The only certainty is that these deviations will occur.

2.4 ENCODE PROJECT

The Encyclopedia of DNA Elements (ENCODE) project was launched in 2003 by the US National Human Genome Research Institute (NHGRI), the Wellcome Trust (UK), Riken (Japan), and the Universities of Geneva and Lausanne (Switzerland). The goal was to determine all of the functional elements of the human genome, including the genes, transcripts, transcriptional regulatory elements, chromatin states, and methylation patterns. All of the data generated from the different institutions, including gene annotation, transcriptome analysis, chromatin analysis, transcription factor binding, and methylation, undergo a verification and metadata process that ensures data integrity and usability. A massive amount of data from this project is now freely available online (http://genome.ucsc.edu/ENCODE/) with multiple search features. Many of the suggestions made by Collins and colleagues have been wisely applied to the ENCODE project, namely international involvement, clear goals, and staging. We will revisit these concepts in Chapter 7, but it warrants repeating, large-scale science can be performed in a scientifically rigorous manner. Proponents of the exposome should heed this advice.

2.5 EPIGENETICS: A CLEAR GENE–ENVIRONMENT INTERFACE

The epigenome will also be addressed in Chapter 3, but an introduction to the topic of epigenetics is appropriate here. While it was initially thought that the primary sequence of the DNA was the driver of heredity, it was quickly discovered that there were additional levels of regulation above that of the primary sequence. These modifications of DNA generally fall under the domain of the field of epigenetics/epigenomics. Epigenetics refers to all of the heritable changes in gene expression that are not coded in the DNA sequence itself, but result in an altered phenotype without changing the genotype. There are three primary mechanisms of epigenetic regulation currently known: DNA methylation, histone modification, and noncoding RNA-mediated silencing. Epigenetic regulation represents an intermediate process that imprints information from environmental experiences on the genome. Thus, the apparently fixed genome can be modified in a way that results in stable alterations in the phenotype (Figure 2.4).

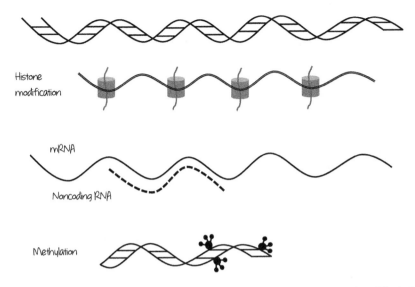

Figure 2.4 Mechanisms of epigenetic modification. Even with the same primary sequence genes can differ in their ability to be transcribed via epigenetic alterations. Histone modification alters the accessibility of the genes to the transcriptional machinery. Noncoding RNA can block regions of mRNA essentially blocking its translation (the basis of siRNA). Addition of methyl (or similar) groups to the nucleotides at CpG islands can effectively turn on or off a gene. Methylation is especially important in the context of the exposome in that chemical exposure can alter the level of methylation.

CpG sites or islands can be methylated by DNA methyltransferases. Methylation of these sites can result in silencing of the gene by repressing transcription. Aberrant CpG island methylation is reported to occur early in the process of tumorogenesis. Histones can undergo multiple covalent modifications that can regulate transcription including acetylation, methylation, and phosphorylation. While many of these processes are part of our endogenous regulation of DNA, most can be directly impacted by environmental exposures. Thus, epigenetic regulation can be viewed as the most direct interface of our genes and our environment. Geneticists may argue that the epigenome is part of the genome and they are correct. But the modification of the genome through epigenetic mechanisms can clearly be considered an external force. The Human Genome Project was focused on the primary sequence. The epigenome is beyond that. It is part of the exposome.

Work from the laboratory of Randy Jirtle was instrumental in showing that environmental factors could induce epigenetic modifications. In the widely cited example using the agouti gene, a gene that encodes coat color in mice, his laboratory showed that modifying the levels of methyl donors in diet could completely change the coat color of a mouse via methylation of the agouti locus. While this was a visually powerful demonstration it was a more powerful mechanistic finding. The genetic sequences of the mice were the same, but their appearance could be transformed through epigenetic modification, and more importantly, inherited. Subsequent work from the Jirtle laboratory showed that the environmental contaminant, bisphenol A, caused hypomethylation of particular CpG islands, which could be reversed by enriching the diet in methyl donors (e.g., folic acid). Moreover, work from the laboratory of Moshe Szyf demonstrated that the maternal grooming patterns of mice could impact the level of methylation in the offspring. Subsequent research has shown that child abuse and neglect and recreational drug use can confer epigenetic changes to our genome. One should shudder at the possible epigenetic alterations that may occur upon exposure to more overtly toxic substances. There is evidence of major environmental exposures leading to transgenerational health effects that are likely due to such epigenetic changes.

Another intriguing level of DNA regulation, which may have an impact on exposome research, is the chromatid end capping by telomeres. Telomeres consist of repeats of a six nucleotide motif (TTAGGG) that act to protect the chromosome from shortening during replication. Degradation and the resultant shortening of the telomeres is correlated with premature aging of the chromosome and an increased incidence of cancer and degenerative disorders. Elizabeth Blackburn, Carol Greider, and Jack Szostak shared the Nobel prize in 2009 for the discovery of telomerase, the enzyme that adds the DNA capping structures. It has also been posited that the telomere length may be an indicator of cumulative DNA damage, which could become a useful component of an exposome index (see Chapter 7).

I must reiterate, discoveries in the field of genetics have been extraordinary. In fact, as scientists learn more about the regulation of DNA and studies based on deep sequencing or whole genome sequencing are conducted, we will likely find that genetics will unveil many more secrets. Indeed, the title of this chapter (genome falls short) may be partially disproven as the science progress. In the future our genes may explain even more than they do now, but our intuition tells us there is a limit. It is important for us to learn what those limits are because this is where the interface with the environment will be critical. Everything beyond this hypothetical limit is by definition nongenetic, that is, exposome.

2.6 OBSTACLES AND OPPORTUNITIES

The somewhat provocative title of this chapter notwithstanding, it is important for exposome advocates to embrace the human genome. The tools, approaches, and strategies from the Human Genome Project and subsequent genetic endeavors will be extremely useful in understanding the impact of the environment in human health and disease. It must be a collaborative venture where genetics is used to provide a biological foundation. Environmental health scientists must continue to elevate the importance and awareness of the environmental influences on health without disparaging the genetic influences. It is not us versus them. Our genes do matter. We have and will continue to reap dividends from the study of our genome, but like many scientific endeavors that work their way into the public sphere there has been a great deal of overpromising. The scientists and politicians overpromise and the public overexpects. The promises of the Human Genome Project have fallen far short of expectations. But even if we had complete mastery of our genome, we would still be missing the external, exposome-like forces that I posit will ultimately account for more than 50% of disease incidence.

The exposome has a lot to learn from the Human Genome Project. One of the most pertinent lessons is the need to have a clear goal. This is one of the major challenges currently facing exposome research. Sequencing the entire human genome was relatively straightforward. There was widespread agreement among scientists in the field that this was a worthwhile endeavor. Even so there was considerable disagreement on the feasibility and approach. The exposome community must be prepared for a much greater level of skepticism and distraction as it is more complex and there is not agreement on what it even is. Other lessons learned from the Human Genome Project that will be useful to the exposome community will be addressed in Chapter 7.

2.7 DISCUSSION QUESTIONS

Companies such as 23andMe can provide extensive genetic information on an individual and it is only a matter of time before personal genome sequencing is a reality. If cost was not an issue would you want to have your genome sequenced? Is there any information you would rather not know?

How do we protect this type of very invasive and private information? Should our health care and insurance providers have access to it?

How does having access to such information change one's behavior? For example, if you know that you had a twofold increased risk of having a heart attack would you alter your behavior?

FURTHER READING

Dolinoy DC, Huang D, Jirtle RL. Maternal nutrient supplementation counteracts bisphenol A-induced DNA hypomethylation in early development. Proc. Natl. Acad. Sci. 2007;104 (32):13056−61.

Franklin RE, Gosling RG. Molecular configuration in sodium thymonucleate. Nature 1953;171:740−1.

Klein C, Lohmann K, Ziegler A. Viewpoint: the promise and limitations of genome-wide association studies. J. Am. Med. Assoc. 2012;308(18):1867−8.

Watson JD. The Double Helix: A Personal Account of the Discovery of DNA. New York, NY: Touchstone; 1986 (Required reading for any scientist. A lesson that scientific discovery cannot be predicted or avoid human foibles).

Watson JD, Crick F. A structure for deoxyribose nucleic acid. Nature 1953;171:737−8.

Watson JD, Crick FHC. Genetic implications of the structure of deoxyribonucleic acid. Nature 1953;171:964−7 (The follow up paper that proposed the copying mechanism of DNA)

Wilkins MHF, Stokes AR, Wilson HR. Molecular structure of deoxypentose nucleic acids. Nature 1953;171:738−40.

CHAPTER 3

The Explosion of -Omic-Based Technology and its Impact on the Exposome

3.1 THE SCIENCE OF "ME, TOO"

Several fields have adopted the term -omic to describe the large-scale approaches to their particular field (genomics, epigenomics, metabolomics, proteomics, etc.). A quick PubMed search for the various -omes and -omic technologies reveals a breadth of acceptance and utility. The genome is obviously the most commonly used -ome and was coined nearly a century ago. Progress on the Human Genome Project spurred similar aspirations for areas examining gene transcription and the translation of proteins. A comprehensive analysis of the complement of genes, transcripts, and proteins was bound to occur. Similarly, understanding all of the ways that the genome could be modified via epigenetic mechanisms was another logical step. After these core -omes were described and corresponding projects initiated, several other fields jumped into the fray. Evaluating one's response to a drug, an exogenous chemical, based on their genomic profile gave birth to pharmacogenomics. Given the close relationship between pharmacology and toxicology, the development of toxicogenomics was another logical extension (more on this in Chapter 4). More recently, the metabolome and microbiome have gained attention and recognition, along with a slew of other -omes, such as the connectome, which focuses on the connections among proteins, arguably a subset of the proteome, the lipidome (the profile of body lipids), the infectome (triggers of autoimmunity), and the phenome (set of all phenotypes within an organism or species). For dozens more examples see http://omics.org/index.php/Omics_classification. Any effort to characterize our exposures at a similar scale will benefit from a close examination of many of these other -omes.

As given in Table 3.1, the genome is the king of the -omes. A PubMed search of keywords highlights this. Genome weighs in with 857,443 hits, proteome with 27,875, and transcriptome with 14,826. Pharmacogenomics (pharmacogenome is not a term that is employed)

Table 3.1 Omics, Omics, Omics. The Major Omic Technologies and their Usage				
-ome (Year Coined)	No. of Citations	Root Term	No. of Citations	Fold-Difference
Genome (1920)	857,443	Gene	1,787,215	21
Proteome (1994)	27,875	Protein	5,278,564	189
Transcriptome (1997)	14,826	Transcript	58,146	3.9
Epigenomics (1950s)	1,214	Epigenetics	6,723	5.5
Toxicogenomics[a] (1999)	1,117	Toxin	273,954	245.3
Exposome (2005)	54	Exposure[b]	583,775	10,811

A search of PubMed for the -ome term and its corresponding root word was conducted. The ratio between the -ome term and the root term gives a sense of the acceptance and use of the -ome term. Genome and proteome are well represented, while toxicogenomics and exposome are not. There is a strong denominator effect here in that there were only 54 citations of the work exposome as of July, 2013.
[a]Toxicogenomics, like pharmacogenomics, never used the -ome suffix, only the -omic suffix.
[b]Substituting environment (1,101,485) for exposure yields a 20,398-fold differential.

has 14,422 entries with metabolome at 3,167 and microbiome at 3,993. Epigenome has 1,214 (epigenetics has 6,723). The connectome has 226 and phenome has 223. When one moves into the environmental health sphere, we have toxicogenomics at 1,117, toxome at 4, and exposome with only 54. Interestingly, the word gene yields 1,787,215, which is only twice that of genome. Transcript gives 58,146, approximately four times higher than transcriptome. Protein yields 5,278,564, which is 189 times higher than proteome. Environment has 1,101,485, which is 20,000 times higher than exposome. What gives? Environment likely appears many times due to a more general use of the term environment, but still the amount of entries for exposome (the denominator) is paltry, likely a combination of its relative youth and slow adoption. Below, the major -omes will be summarized, highlighting some of the important technical approaches that have contributed to the discipline.

3.2 TRANSCRIPTOME

The ability to print thousands of nanoliter drops of complementary DNA (cDNA) onto a glass slide was a key step in the development of the microarray and the ability to measure thousands of transcripts in a single sample. Hybridization of mRNA to cDNA was a mainstay in molecular biology; it was just necessary to miniaturize the process. The laboratories of Pat Brown and David Botstein employed simple robotics and fountain pen technology to generate the first microarrays. After mRNA is isolated from a biological sample, a more stable sample of cDNA can be generated through the process of reverse transcription. By labeling two different populations with unique fluorophores, it was possible to measure the relative abundance of the bound sample, corresponding to a particular cDNA. Soon after, shorter oligonucleotides with overlapping sequences were spotted on the arrays, which served as the basis of the Affymetrix platform. This technology avoided the need to keep large stocks of cDNAs, and because it used multiple overlapping oligonucleotides it was able to detect changes at the single nucleotide level. The cDNA approach did not have this level of specificity. The oligonucleotide-based approach also allowed the assessment of up to two million SNPs. As these approaches attempted to measure all of the transcripts, the results were often referred to as the transcriptome. More recently, high-throughput sequencing of RNA (RNA-seq or whole transcriptome shotgun sequencing) has been employed to provide a higher level of resolution. The potential benefits of RNA-seq are the more complete coverage of the more than 10,000,000 SNPs in the human genome, the possibility of direct sequencing of mRNA without the need for reverse transcriptase, which can introduce errors in the sequence, and sequencing of non-mRNA species, such as noncoding RNA. The transcriptome will continue to advance as newer methods provide greater resolution and fidelity.

3.3 PROTEOME

Proteomics has significant advantages over transcriptomics because it is measuring the actual effector molecules in the system, that is, proteins. One must infer that the altered gene transcription will ultimately lead to a change in protein expression, whereas proteomics is directly measuring the protein. Assessment of the entire complement of proteins in a cell or an organism is the goal of proteomics. Proteomics assesses which proteins are present, their relative abundance, and various modifications to the proteins. The separation of proteins based on size and charge is routinely performed using acrylamide gels. In fact, the initial proteomic studies were based on two-dimensional (2D) separation by size and isoelectric points. The various spots on the gel could be visualized by silver staining and the size of the points correlated to the relative abundance. It was possible to transfer the 2D gel to nitrocellulose and perform immunoblotting to identify proteins. In addition, by excising a particular spot and subjecting the purified sample to digestion and mass spectrometry it was possible to identify the protein by the unique mass-to-charge ratios of the cleaved proteins using matrix-assisted laser desorption/ionization time-of-flight (MALDI-TOF) or electrospray ionization with time-of-flight (ESI-TOF) mass spectrometry. For example, a 70 kDa protein has approximately 60 amino acids. The trypsin digestion breaks it up into peptides with an average molecular mass of 1,000 daltons, about 9–10 amino acids. Thus, our sample protein is cleaved into six or seven pieces and that pattern of fragments, when compared to the reference database and reassembled computationally, allows identification of the protein. Even with a one-dimensional separation by mass, it is possible to distinguish among many proteins sharing the same total mass, as their trypsin fragmentation patterns are unique. The peptide fragmentation approach is laborious, but is an excellent way to identify proteins. However, it is not optimal for getting full coverage of the proteome. With refinement of the fundamental steps of the separation of proteins followed by identification by mass spectrometry, higher throughput technologies have become available.

Separation of tens of thousands of proteins in a given sample does not have to be conducted on an acrylamide gel. Ingenious protein chemists developed methods to separate the proteins with combinations of ESI or MALDI and tandem mass spectrometry (MS/MS because the mass spectrometry occurs in sequence with some sort of fragmentation in between) or Fourier transform ion cyclotron (FT-MS). Instead of starting with cleaved peptides, the more recent approaches start with samples containing intact proteins. The proteins are fragmented during the process but not with enzymes. These newer techniques are more expensive, but more amenable to high-throughput proteomic experiments.

The multiple levels of modification and the temporal variability in expression make the proteome much more complicated than the stable genome. The identification of proteins is an important outcome of proteomics, but the ability of the proteomic approaches to identify protein modifications may be one of the field's strongest features. There are numerous ways proteins can be modified including glycosylation (carbohydrate/sugars), phosphorylation, ubiquitination (a marker for degradation), and sumoylation (small ubiquitin-like modifier that impacts localization and stability). These posttranslational modifications prime the protein for degradation via autophagic mechanisms (ubiquitination), activate enzyme systems (phosphorylation), and are involved in the maturation of proteins (glycosylation). Other forms of posttranslational modification include nitrosylation, oxidation, methylation, and acetylation. Oxidation was once thought to be a mere by-product of injury, but the reduced/oxidative state of proteins may serve a regulatory role for many proteins. Various initiatives to characterize the human proteome are underway. The Human Proteome Organization (HUPO) consists of two parallel projects. One is the chromosomal-based human proteome project (C-HPP), with one research team assigned to each chromosome, and the other is the biology/disease-driven HPP (B/D-HPP), which will examine specific disease states and biological entities or processes. The project maintains a website that provides descriptions and updates of the project (http://www.thehpp.org).

3.4 PHARMACOGENOMICS/TOXICOGENOMICS

There is evidence that people with certain genetic polymorphisms respond differently to certain drugs. Pharmacogenomics attempts to identify these sensitizing genes and then test them in potential users. For the most part, these are specific genes that alter the susceptibility to a particular drug, either through enhanced or impaired metabolism or altered expression of the target molecule. There are many drugs that only work in individuals with a particular genetic profile. Many in the field believe that pharmacogenomics will become an integral part of personalized medicine by tailoring drug treatments to a person's genetic background. Toxicogenomics was conceived as a way to take advantage of microarray technology to measure the impact of toxic chemicals. Etymologically, it should have been toxicotranscriptomics, as it was using microarrays to assess the effects of toxic chemicals on gene expression. More recently, some in the field have proposed that toxicogenomics should include any -omic data that assesses the toxicological consequences (more detail in Chapter 4). If one accepts the term to encompass all -omic technologies that inform the field of toxicology then it clearly falls under the larger umbrella of the exposome.

3.5 EPIGENOMICS

By examining the epigenetic modifications discussed in Chapter 2 at a genome-wide scale, epigenomics provides a global measure of altered DNA regulation. In the US, NIH has established a Human Epigenomics Mapping Consortium to serve as a resource for investigators. Taking advantage of next-generation sequencing and other advances, the consortium will generate and share data on DNA methylation, chromatin accessibility, histone modification, and small RNA transcripts at the genome-wide level. The ENCODE project, mentioned in Chapter 2, is serving as a repository for these epigenomic data sets.

3.6 METABOLOME

Metabolomics is the study of the chemical metabolites that result from cellular processes. Many of these metabolites are readily found in the bloodstream, providing an accessible source for sampling. By examining the state of the chemical processes within a cell or an organism, metabolomics provides information on the physiological state of the cell. For example, a transcriptomic or proteomic analysis may tell us that a particular enzymatic pathway is upregulated in a particular organ, but it does not tell us if the substrate is present, nor if the product is being formed. By providing information on the substrates and products, it is possible to determine what activities are occurring in the cells and quickly infer which proteins are active. Metabolomics shares many features of proteomics, in that one is attempting to measure the entire complement of the target entity (protein vs metabolites) in a complex mixture. The challenges of separation and identification are quite similar. The major difference is that the metabolites tend to be smaller molecules than proteins (molar mass of glucose is 180 g/mol, amino acids average ~110 g/mol, but proteins can be composed of hundreds to thousands of amino acids).

David Wishart at the University of Alberta reported the first draft of the human metabolome to contain 2,500 metabolites, 1,200 pharmaceuticals, and 3,500 dietary constituents (see Human Metabolome Database, www.hmdb.ca). However, when one considers that plants also contain thousands of metabolites, ingestion of a diverse diet could greatly increase these numbers. It is also likely that many of the chemicals classified under pharmaceuticals are actually environmental contaminants. Similar to proteomics, a variety of mass spectrometry-based approaches have been used in metabolomics, including gas or liquid chromatography combined with mass spectrometry, tandem mass spectrometry, and Fourier transform mass spectrometry. Nuclear magnetic resonance has also been used for metabolomic analysis. The goal is to be able to resolve as many of the metabolites as possible without sacrificing sensitivity. Advances in separation, detection, and bioinformatics will likely increase the sensitivity and resolution, leading to an even greater number of detected metabolites. It will be especially attractive to be able to layer these metabolomic patterns onto the proteomic and transcriptomic patterns to see how well they correlate. Such analyses will require systems and computational biology approaches that will be discussed in Chapter 5.

3.7 MICROBIOME

The microbiome consists of the millions of microbial organisms that exist within the confines of the human body. Primary among these are those that reside in the intestines of the digestive system, but microbes also reside within our urogenital systems, the integumentary system, and also in our noses and mouths. These species include bacteria, viruses, and single-cell eukaryotes. The gut microflora has been somewhat underappreciated in the past, although most people acknowledge their importance after a course of strong antibiotics that can decimate these important microorganisms. The human microbiome is thought to contain more than 10 times the total number of cells found in the human body and due to their much smaller size contribute 1−5 lb of human body weight. With the diversity of organisms, there are likely 100 times more genes in the microbiome than in the human genome. There is potential for extensive metabolism of the nutrients and chemicals that enter our intestines. Harmful substances may be metabolized to less harmful species. Conversely, somewhat inert compounds may be bioactivated by the microbes to more harmful substances. These microbes can also influence absorption of nutrients and toxicants across the gut. This may be deleterious if nutrients are being blocked, but beneficial if harmful xenobiotics are being impeded. One of the advantages of microbiome research is that it completely builds upon the success of genomic technologies by simply sequencing the genomes of the microorganisms that reside within our bodies. We will likely be learning more about the role of the microbiome in health in the coming years. It is somewhat staggering that an appreciation of the role of the trillions of foreign cells in our bodies only developed about 20 years ago. It is possible that much of what we consider to be interindividual variability in response to exposures or drug treatments may be due to an individual's microbiome. Indeed, some conditions are even treated by repopulating the microbiome with specific microorganisms. It is important to consider the impact of the microbiome on the exposome. For example, an approach that only looks at the exogenous exposures without considering what happens to the chemicals once they get inside our body would totally miss the effects of the microbiome (Figure 3.1).

Fortunately, the evaluation of chemicals in our bloodstream may somewhat obviate a need to specifically measure many aspects of the microbiome, since many of the processes in our bloodstream are occurring downstream from the effects of the microbiome. For example, if disruption of the microbiome reduced absorption of a particular nutrient by 50%, then one would expect to see much less of the nutrient in the bloodstream. It is possible that much of the variation that is observed among human metabolomes is the result of differential microbiomes. Such experiments could be readily conducted using humanized mouse models, in which the human microbiome is transferred into the mouse.

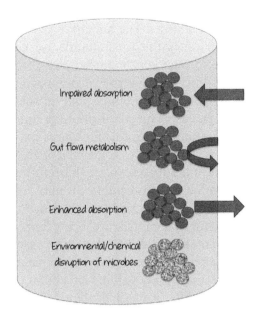

Figure 3.1 The microbiome−exposome interface. The microbes that reside in our digestive, urogenital, and integumentary systems have a significant impact on our health. As shown here, bacterial colonies can alter absorption of nutrients or toxicants from the gut, metabolize nutrients or toxicants (degrade or activate) before they are absorbed into our bloodstream, and environmental chemicals can disrupt the balance of the microbial colonies. The role of the microbiome in the body's response to environmental exposures is an important consideration for the exposome.

3.8 WHAT TO DO WITH ALL OF THESE -OMES

While it may be a bit premature to make this case, one could argue that the aforementioned -omes are actually all part of, contribute to, or are tools of, the exposome. If a person was placed into a vacuum, their genetic code (genome) would dictate the composition of their transcriptome and proteome, but once they exit the vacuum these components start responding to the environment. The subsequent alterations in gene and protein expression, epigenetic changes, metabolic alterations, and changes in the microbiome all are part of the exposome. The responses are either genetically encoded or environmentally imbued. It is either genome or exposome (Figure 3.2).

3.9 WHAT TYPE OF OMIC IS THE EXPOSOME?

I have intentionally not used the term "exposomics." Omics imply a unique approach. In general -omics refers to a particular field of study,

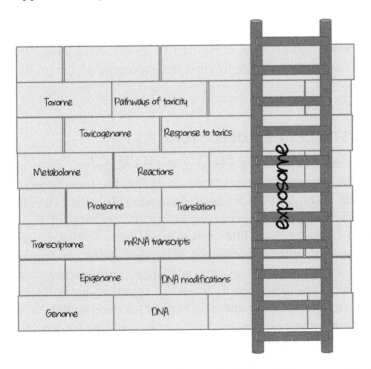

Figure 3.2 The wall of -omes. The various -omes are shown as levels in the wall. The genome is at the base followed by the -omes representing the gene modifications, expression level, proteins, metabolites, and pathways. The exposome will need to travel up and down the wall to collect the key information from the existing -omes.

while the -ome refers to the object being studied. The exposome is an entity hoping to be defined. It is the object of study, but the techniques and approaches being used to study it already exist in a variety of fields. It could be argued that no new techniques will be needed to define the exposome; thus, there may not be a need for a new field of exposomics. I believe that the field studying the exposome has been and will continue to be environmental health sciences. The exposome is just one particular construct that is primarily within the domain of environmental health sciences.

I am not ruling out a new field of exposomics or the use of the term. It may be that the assessment of the exposome will require a host of new tools and unique approaches that constitute a new field; however, at this point it is not clear what that suite of tools would be and it may be premature to tout a term that is to date ill-defined. A human exposome project will surely require advancement in technology, but this could be an advancement in LC-MS, bioinformatics, or computational biology; technology that is not exclusive to the exposome. The integration of the myriad of data from existing disciplines may require some unique approaches, but the study of complex biological processes and systems is referred to as systems biology (more on this in Chapter 5). I prefer to view the exposome as the goal, not the process.

3.10 OBSTACLES AND OPPORTUNITIES

One must be careful when adopting -omic vocabulary. Over one hundred -omic terms have been coined, including the intentionally absurd ridiculome   (http://www.biomedcentral.com/1741-7007/10/92).   This -omic abuse has, not surprisingly, led to the creation of the hashtag badomics (# badomics). The explosion of -omes and -omic terms has been somewhat unfortunate in the biomedical sciences. The idea of a comprehensive evaluation of certain biological data sets is admirable, but coining terms to artificially boost the apparent importance of one's topic of study is not desirable. It is critical for investigators in environmental health sciences to avoid such a perceived fate for the exposome. The field has struggled with how to deal with complex exposures, especially as it relates to disease causation/contribution and the development of regulatory guidelines. While many have criticized the concept of the exposome (or at least the word), few would summarily dismiss the idea that the environment has an impact on human health, and that these exposures represent complex mixtures. Thus, we must be methodical and careful in our approach to define the exposome. We must focus on verifiable outcomes that have the potential to impact human health. We also want to take advantage of the data and tools resulting from other relevant and validated omic-based technologies. While comparisons with the genome, transcriptome, and proteome are welcome, we would do best to avoid comparison with the ridiculome.

3.11 DISCUSSION QUESTIONS

If study of the human genome gave rise to genomics, does the human exposome give rise to a new field of exposomics? Should it be a separate field or should the exposome serve as an organizing framework for the cornucopia of -omic technologies?

Put yourself in the position of arbiter of -omes. You must decide which 12 omes/omics remain in the scientific domain and which are stricken from the scientific record. Take note of your criteria for inclusion or exclusion. What are your choices?

FURTHER READING

Aebersold R, Bader GD, Edwards AM, van Eyk JE, Kussmann M, Qin J, et al. The biology/disease-driven human proteome project (B/D-HPP): enabling protein research for the life sciences community. J. Proteome Res. 2013;12:23–7.

Caron H, van Schaik B, van der Mee M, Baas F, Riggins G, van Sluis P, et al. The human transcriptome map: clustering of highly expressed genes in chromosomal domains. Science 2001;291 (5507):1289–92.

The Human Microbiome Project Consortium. Structure, function and diversity of the healthy human microbiome. Nature 2012;486:207–14.

CHAPTER 4

The Exposome in Environmental Health Sciences and Related Disciplines

4.1 WELCOME HOME EXPOSOME

It is clear that the intellectual home of the exposome is the field of environmental health sciences. The field focuses on how our environment influences health and disease, and the exposome has the potential to greatly improve our understanding of these effects. In general, environmental health sciences encompasses research conducted at the population level, assessment of actual environmental exposures, and toxicology, which explores the biological basis of the associations observed at the population level, as well as the mechanisms by which the measured exposures alter biological systems. There are other components of environmental health sciences, such as risk assessment and disease ecology, both of which fit nicely into the exposome paradigm (Figure 4.1).

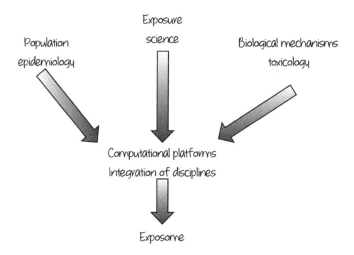

Figure 4.1 The components of environmental health sciences and the exposome. Each of the subdisciplines of environmental health sciences can contribute to the elucidation of the exposome. Information and findings from populations (via environmental epidemiology), accurate measurement of chemical species in our surrounding media (exposure science), and data on biological mechanisms and target pathways (via toxicology and biochemistry) will all be needed for exposome research. The findings from the three major subdisciplines need to be integrated via computational methods, along with data from atmospheric chemistry, genomics, behavioral science, nutritional sciences, and many others.

Environmental health sciences is at its best when it capitalizes on the transfer of knowledge among the three major divisions within the field. Toxicologists rely on epidemiological studies and analysis of exposures to determine which compounds should be studied. Epidemiologists rely on toxicologists to determine if their observed associations are in line with what is known about the biological and toxicological pathways involved in the health effects. Toxicologists also look to the exposure scientist to know what the relevant exposure levels are. When data from all three subdisciplines converge it provides a very strong scientific basis for that particular environmental exposure having an impact on human health. A similar level of convergence among all components of environmental health sciences will be necessary for data from the exposome to have an impact on human health (Figure 4.2).

One of the challenges for the field is that, historically, environmental exposure studies are those that are unintentional. Ingestion of drugs or alcohol, cigarette smoking, overeating, and levels of physical activity typically fall outside the domain of the field. Separating exposures that have a volitional component from those that are passive is a reasonable structure. But exclusion of exposures from the occupational setting is somewhat troubling for the individual. When

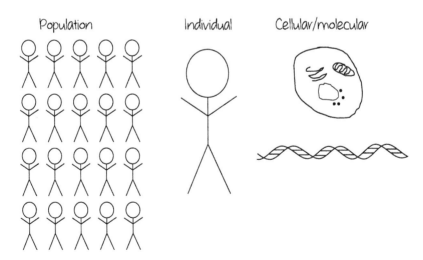

Figure 4.2 Populations to molecules. Exposome-related data will come from a broad range of sources, including population studies, personal environmental and biological monitoring, and from laboratory studies of animals, cells, and molecules.

occupation is removed from the definition of environment, one of the biggest sources of potential exposures is being arbitrarily excluded. Part of this classification structure is also due to the organizational structure of the agencies that fund this research. Within the US NIH, drugs of abuse fall under the National Institute of Drug Abuse, alcohol consumption under the National Institute of Alcoholism and Alcohol Abuse, while obesity and physical activity have no specific home and are addressed by multiple branches. Regardless, a focus on unintentional environmental exposures is the charge of the field of environmental health sciences, but the field must be receptive to other exposures that may be due to voluntary activities or occupations. One way to think of it is that the field of environmental health sciences should focus on the *measurement* of classical environmental exposures, but be *inclusive* when it comes to other types of data when interpreting the former. This is because our environmental exposures occur within the context of these other dietary and lifestyle exposures, and ignoring these coincident exposures could seriously confound results. For example, a community that resides near a factory may have a higher incidence of a particular disease and if one only focuses on those potential exposures from the factory and does not take into consideration the various socioeconomic, dietary, and lifestyle factors, the true drivers of the disease could be missed. An elevation in the lipid soluble chemical could well occur, but the reason for the increased disease risk could be because of the consumption of a high fat diet and not necessarily the concomitant increase in the chemical. The chemical may just be a biomarker of an unhealthy diet. Of course, environmental epidemiologists control for most of the socioeconomic and lifestyle factors, but it is extremely difficult to control for complex biochemical factors that include variations in cytochrome p450, redox pathways, and fluctuations in hormonal systems. For the most part they are controlled using questionnaire and census-type data, not biological data. Having parallel laboratory analysis to assess nutritional and redox status, confirm smoking status, and measure genetic vulnerability would be beneficial in determining the appropriateness of the control group. Further, with the availability of improved measures of exposure and biological response the traditional survey instruments can be refined and strengthened to improve their predictive value. Ideally, we focus our attention on those chemicals that do have an adverse health effect and work to remediate them.

Environmental health scientists must be the drivers of the exposome. Investigators from other fields have been more receptive to the concept than many of those within the field. It is not to say that we abandon our traditional approaches, in fact, we probably need to conduct more of the traditional type of research. But the exposome demands that we draw data from outside our historically defined intellectual boundaries. We do not have to measure all of these other aspects; however, we need to include them in the interpretation of our data. As it turns out, many of our analytical approaches can and do measure the variables that lie outside of environmental health, so why not include them in our analyses?

4.2 TOXICOLOGY—MECHANISMS OF TOXICITY

The concept of the exposome was essentially laid at the feet of exposure assessors and epidemiologists. The argument was that epidemiologists needed better and more comprehensive data from the exposure assessors. It was up to the field of exposure assessment/science to deliver the goods. There was no obvious place for the toxicologist. But if one accepts the more expansive definition of the exposome, that is, representative of all external forces, not just exposures *per se*, and the body's responses, the role of the toxicologist becomes readily apparent. The mechanistic understanding of environmental agents on biological systems becomes foundational.

Toxicology is defined as the study of the adverse actions of chemicals on biological systems. This is a very broad definition that includes every type of chemical and every type of biological system, from plants, to bacteria, to fish, to humans. For the purpose of this book though, we will focus the discussion of toxicology on the adverse effects of chemicals on human biology. The importance of toxicology to exposome research is that it provides information on the biological effects of toxicants. Toxicologists determine how chemicals interact with organ systems, specific pathways, cells, and macromolecules.

Toxicology, as a field, should adopt the exposome as a means of integrating their mechanistic data into human health paradigms. This is not to say that toxicologists need to dramatically change what they are doing, but rather they need to position themselves to be able to contribute their findings to the exposome paradigm. The biological pathways affected by the toxic compounds are a key component of exposome-based research. Dr. Thomas Hartung at Johns Hopkins University has suggested that a Human Toxome should serve as a framework for study within the field of toxicology. The Human Toxome is analogous to the exposome, but with a specific focus on the adverse effects of toxic environmental chemicals. The findings from such an effort should easily be incorporated into the exposome framework (this academic project should not be confused with the Human Toxome Project being conducted by the Environmental Working Group (EWG) and Commonweal in California, which is focused on measuring particular chemicals in a relatively small number of people. The apparent goal of the EWG initiative is to raise awareness about the presence of these chemicals, but due to its limited experimental design, it is unlikely that the data generated will be of value to the exposome efforts or the scientific community at large). One could argue that Hartung's narrower approach is more in line with the traditional view of environmental exposures. The type of data collected by this initiative, especially the identification of pathways of toxicity (PoTs), are laying the foundation for a major step forward for toxicological risk assessment, but they could also be essential for the interpretation of exposome data. The toxome approach and the Tox21 initiative (described below) should be encouraged, in that they are generating data that have the potential to add to the body of the exposome. It will be important to pursue these studies with the mindset that the findings can be absorbed into the exposome framework. For example, if a toxome approach identified 25 different pathways involved in carcinogenesis it should be possible to build a computational model that allows screening of combinations of chemicals for their ability to disrupt these pathways and contribute to regulatory risk assessment. The computational nodes could become integrated into a larger exposome model and data from exposome efforts could be used to identify the combinations of chemicals to be tested in the PoTs identified in the toxome efforts.

Having been trained as a toxicologist, I have an admitted bias toward toxicology and the study of biological mechanisms of actions. From the use of curare to understand the neuromuscular junction, to the work of the late Toshio Narahashi using tetrodotoxin to understand sodium channel function, toxins have been used throughout history to uncover biological secrets. Dr. Narahashi went on to use that insight from the naturally occurring tetrodotoxin to determine the actions of the pesticide DDT and many other synthetic chemicals. Learning about biology by paying attention to the lessons taught by nature is a recurring theme throughout science. Understanding how the various environmental toxicants alter human biology is critical to understanding disease pathogenesis and can identify molecular targets and pathways that can be targeted for intervention.

The phrase, "the dose makes the poison" is a useful construct for explaining the concept of dose−response. Distilled from the words of Paracelsus, the Father of Toxicology, the phrase elegantly reveals the relationship of the dose of a compound and its effects (adverse in this case). Unfortunately, this toxicological catchphrase is unable to capture the nuances and shape of the curve. Decades ago the relationships were thought to be not far from linear. Paracelsus taught us that good things can be bad at high doses, so the thought that bad things can be good at low doses seems counterintuitive. But in fact there is strong evidence to support the concept of hormesis for many classes of compounds and some biological pathways. Hormesis is the term used to describe the phenomenon where low doses of a compound, which is demonstrably toxic at higher doses, may actually provide a beneficial effect. Low levels of oxidative stress activate complex response pathways that make it easier for the body to respond to subsequent insults. While it is a bit speculative, this could be part of the reason that antioxidants routinely fail in clinical studies of disorders that are thought to involve oxidative stress in their pathogenic process. Low levels of some stressors may be beneficial to the organism. Ed Calabrese has worked tirelessly over the past decade or so to bring the concept of hormesis back into the dose−response discussion. A basic scientist can consider hormesis and look for its qualities in laboratory studies. However, the concept of hormesis can cause apoplexy in the regulatory toxicologist. When making decisions about safe levels of environmental chemicals the idea that there may be a beneficial effect just does not fit into the mental construct of regulatory science. Dr. Calabrese and others have documented the hormetic effects in many situations. It appears that radiation may even have some hormetic effects. This is very challenging to the dogma of any exposure to a mutagen is bad. So what happens when someone is exposed to dozens of chemicals theorized to have a hormetic effect? If the chemicals are related, do the effects become additive pushing the curve past the positive to the adverse? This is where approaches that look at the combination of chemicals to which we are exposed could provide needed insight. The exposome is open to this concept in that much of the work will proceed in an unbiased and hypothesis-free format.

Computational toxicology is a more recent development within the field of toxicology. While most are familiar with the quantitative structure activity relationships (QSAR) used in drug development, a similar approach can be used in toxicology to model potential toxic interactions. The US Environmental Protection Agency (EPA) has established a Computational Toxicology Center to develop and apply computational tools to aid with the prediction of toxicity as it relates to regulatory decision-making. The fact that only a small fraction of chemicals introduced into our environment have undergone toxicological testing highlights the need for faster and less expensive means of assessing toxicity. The goal is that predictive toxicity models can be developed, allowing identification of the compounds with the highest likelihood of being toxic so that they can undergo the more extensive testing. When coupled with recent advances in high-throughput technologies the potential for expanding the number of compounds screened or tested is very high.

The idea of high-throughput toxicology is that we need to gain information on thousands of chemicals, but most classical toxicological studies examine compounds one by one. It is exceedingly difficult and expensive to test compounds in the types of studies required by regulatory agencies. If classes of chemicals could be tested under the same conditions it may be possible to extrapolate potential adverse effects. Many assays on cellular growth or cellular injury can be performed in 96, 384, or 1,436 well trays using robotics. This allows for the high-throughput-type approaches that are being conducted under Tox 21, a joint venture among NIEHS, EPA, the National Center for Toxicology Research at the Food and Drug Administration, and the National Center for Chemical Genomics at NIH. Tox21 will test over 10,000 chemicals on a battery of high-throughput assays, providing data on compounds that have never undergone regulatory analysis. Such data can be used to develop and validate computational models that may permit *in silico* testing of hundreds of thousands of compounds.

Toxicologists who conduct research in the biomedical versus the regulatory side are being challenged by funding agencies to demonstrate the translational nature of their research. For many the connection is not obvious; however, the exposome provides a useful translational conduit. If findings from toxicology studies make their way into the exposome construct, which they should, then by its very nature the research will be translational as it is being incorporated into a model aimed at improving human health.

4.3 EXPOSURE SCIENCE (OR ASSESSMENT, OR BIOLOGY)

For decades, the field of exposure assessment has been determining the composition of our surroundings and how various contaminants enter the body. More recently, the term exposure biology has been used to describe the study of the biological aspects of exposures and is one that biomedical funding agencies seem to prefer in that it is more aligned with their biologically-based approaches. From the standpoint of environmental health sciences and the exposome, the more expansive and inclusive exposure science would appear to be the best descriptor of dealing with exposures. Exposure science not only encompasses the measurement of contaminants in our air and water, but also includes measuring the chemicals within the body or other biomarkers of exposure. For example, the chemical composition of an airborne pollutant and its physicochemical interactions with other airborne chemicals would not necessarily be part of exposure biology, but would clearly be of importance and fall within the domain of exposure science. This is similar for exposure assessment. Assessing what is in the environment is only one component of exposure science. Without an understanding of how chemicals move from the environmental matrix to the target species, it is difficult to predict the biological consequences.

The field of exposure science/assessment was one of the first to embrace the exposome concept. Indeed, Dr. Wild threw down the proverbial gauntlet at the feet of the field of exposure science. In the US Steve Rappaport and Martyn Smith from the University of California at Berkeley were leaders in this area. Working from their NIEHS-funded Superfund Project they proposed that the exposome could advance the field. Dr. Rappaport has also proposed that the measurement of DNA adducts could represent a biological measure of cumulative chemical exposures. Drs. Rappaport and Smith have been key proponents of the exposome concept. Dr. Paul Lioy, a leader in the field of exposure assessment, has suggested that semantic arguments about terminology are misguided and that we are wasting time by not doing the actual science. While I agree that conducting the science is the ultimate goal, the framing of the questions and scope of the problem, which includes definitions, is not trivial. However, I would argue that if it is framed correctly, the exposome has the potential to attract more attention and financial support for the evaluation of the environment in our health. From the perspective of an exposure scientist it is understandable why a focus on internal biological markers of exposures would be irritating. If one believed that the exposome would only rely on internal biological markers there would be a reason for concern. But this is not the case. The exposome, as defined here and as defined by Dr. Wild, incorporates the rich exposure data provided by the field of exposure science, as well as the data collected from human samples. In fact, the environmental exposure data are critical in the interpretation of the internal biological markers of exposure and response.

Dr. Lioy has suggested the use of the term eco-exposome to reinforce the importance our surrounding environment, including the health of the planet. This concept was also introduced in the National Academy of Sciences report on Exposure Science for the 21st Century, which Dr. Lioy cochaired. The working group also proposed the use of the term eco-exposome to explain the exposures that reside outside of the body, but the revised definition of Miller and Jones cited in Chapter 1 is all-encompassing, making the need for such a distinction obsolete. Creating additional permutations to a concept that is already battling intellectual inertia will just weaken the case. Investigators may focus on one specific aspect or subset of the exposome, such as those described as part of the eco-exposome, similar to how a person in the field of proteomics will only study one specific posttranslational modification. But there is not a posttranslational proteome, just a proteome. Similarly, it would be preferable to incorporate all aspects into a unified definition of the exposome, lest we provide more fodder for the -omic critics. Dr. Brunekreef at Utrecht University has also addressed these distinctions, noting that the definition of the exposome put forth by Rappaport and Smith is more narrow versus the more expansive and inclusive view of Wild. Dr. Brunekreef also noted that the eco-exposome broadens the exposome sphere even more. The health of the planet is certainly a key component to our own health, but ecological concerns, such as weather patterns, climate change, and plant health are outside of the working model of the exposome that is focused on human health. A universal definition of the exposome must be adopted.

To illustrate the importance of measuring environmental contaminants outside and inside the body, the reader is asked to examine the soup analogy (Figure 4.3). Imagine tasting the following items individually: raw carrots, raw chicken (remember you are just imagining this, no need to worry about salmonella), raw potatoes, raw onions, salt, and pepper. Now describe the taste of the soup that will be made from these ingredients. It is difficult. What if the amount of onions was doubled, or the pepper was halved? Until you taste the final compilation that has simmered for several hours, you will not know what it will taste like. Assessing the impact of the environment on our health is similar. Measuring the individual components outside of the body cannot tell you what the soup will taste like, no more that tasting the soup can allow you to describe the precise ingredients and quantities used to make the soup itself. In order to truly understand the soup, you must know what goes into the soup *and* what the end result tastes like. The exposome is useless without the combined knowledge of the individual ingredients that go into it and an assessment of the merged ingredients within the biological system. If it is determined that a particular combination of exposures increases the incidence of the disease, knowing the components in our environment can inform the steps to mitigation. To take this cooking analogy a bit further, imagine if a group of 10 people were given the recipe for this soup. Would the end result be the same for all 10? Unlikely. Some would cook it longer, dice the carrots up finer, cut the chicken into different shapes, or use more or less salt. The individual variability in cooking style pales in comparison to the individual variability of our biology. Just because several individuals are exposed to the same ingredients does not mean that they will end up with the same soup. In fact, it is highly unlikely that they will (Figure 4.3).

The implementation of more person-specific exposures and responses, the types of approaches being discussed for the exposome, are necessary to address the individual variability. When we have

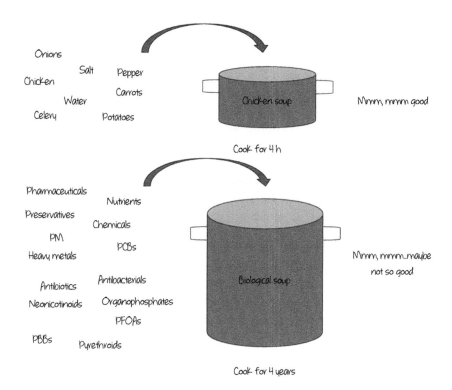

Figure 4.3 The soup analogy. To make soup individual ingredients are collected, prepared, and added into a pot with water and simmered for hours. One can view our bodies (and bloodstream) as a type of soup composed of the various ingredients we add through our exposures. If one samples the individual ingredients of the soup, it may be possible to estimate what the end result will be, but it will be impossible to know exactly how it will taste. Until the ingredients are put in the pot together and are subjected to the cooking process, which involves metabolic conversion and molecular interactions of the ingredients, the ultimate flavor of the soup is not apparent. Likewise, sampling the individual ingredients outside of our bodies cannot tell us what the essence of our exposures is. Sampling the final product will allow us to estimate what ingredients went into it, but not necessarily their forms and amount. However, if one conducts a thorough analysis of the individual ingredients before they are added and then conducts a similarly detailed analysis of the end result it is possible to determine the level of transformation and interaction of the various ingredients after entering the body. A major goal of the exposome is learning how the various ingredients and processes impact the flavor of the soup so that we can make the soup taste better. Note: polychlorinated biphenyls, PCB; particulate matter, PM; Perfluorooctanoic acid, PFOA; polybrominated biphenyls, PBB.

information from personal environmental monitoring and biological measures, which incorporates individual variability, and then combine those data with environmental monitoring station data, a powerful set of information begins to emerge. It becomes possible to determine the relative effects of individual activities, behaviors, and housing, etc., on exposures. Indeed, if such a field developed that integrated individual and population exposure data with biological modifications and health outcomes, I may capitulate on my stand against a new field of exposomics.

4.4 EPIDEMIOLOGY AND THE EXPOSOME

I suspect that the environmental epidemiologists will be the last group to embrace the exposome. The epidemiologists need validated measures of exposures. Hypotheticals are not any better than the real measures that exist today. In fact, I would be concerned if this group readily embraced the exposome without appropriate validation. It is up to the exposure scientists, toxicologists, and others in the biological arena to demonstrate that an exposome-type measure is truly measuring an outcome of biological importance. Once this is established then it will be incumbent upon the epidemiologist to use the measures that provide more detailed information about exposures and responses.

Enhanced quantitative information on exposures will improve the ability of epidemiological studies to identify the underlying cause of disease. The exposome is ultimately an epidemiological construct. The textbook definition of epidemiology is "a branch of medical science that deals with the incidence, distribution, and control of disease within a population" or "the sum of the factors controlling the presence or absence of a disease or pathogen." Environmental epidemiology focuses on the environmental factors that contribute to the incidence, distribution, and control of disease within a population. Thus, the ascertainment of the environmental factors is essential. To date, there has not been a means of capturing the impact of complex mixtures, which is something the exposome can address.

The majority of epidemiological studies focus on infectious agents. Such agents are amenable to studies of transmission and spread of disease. The nature of infectious agents is more tangible, in that the pathogens can be identified via PCR, sequencing, or ELISA, and their vectors can be measured and tracked. While some environmental factors are somewhat easy to assess, many others are much more difficult. For example, determining one's smoking history is fairly straightforward. However, determining one's exposure to a particular pesticide can be daunting. Individuals do not perform well when recalling their past exposures, and interviews and survey instruments only address the potential exposures, not the actual biological exposures. This is where the field of exposure science has had a major role and where Dr. Wild encouraged improvements. By directly measuring the chemicals that are in an individual's or a population's environment it is possible to estimate the exposure. The measurement of what is the environmental matrix is highly accurate. However, translating the quantity of chemical in the environment to what ultimately ends up in a person's body is much less definite. Indeed, the individual variability in uptake, absorption, distribution, metabolism, and excretion is vast. This is where the exposome can have a major impact. By measuring the compilation of chemicals in the human body and comparing that to the concentrations in the surrounding environment one can start to address the issue of individual variability and determine what the most accurate predictors of chemical deposition are in an individual's bloodstream.

4.5 GLOBAL EPIDEMIOLOGY

In the landmark 2010 Global Burden of Disease report, over a dozen risk factors are shown with their relative impact on disease. It must be noted that this is a global evaluation. If one focused on one specific country there would be notable shifts in many of these factors. Historically, indoor air pollution and ambient particulate matter would have been addressed by environmental health scientists, but most of the others would be considered to be outside their domain. However, the exposome captures all of the risk factors addressed in the Global Burden of Disease project. Smoking, alcohol consumption, and low physical activity are part of the lifestyle or behavior component. High blood pressure, high body mass index, high fasting glucose, and high cholesterol can all be viewed as part of the cumulative biological response. Low fruit intake, low nut and seed intake, high sodium, and iron deficiency are all captured under dietary influences. All of these risk factors are part of the refined definition of the exposome. They are the modifiable components that could well be part of a comprehensive exposome index.

4.6 PESTICIDES AS AN EXAMPLE OF STRAINED RELATIONSHIPS

Insecticides, herbicides, and fungicides are used throughout the world to control insect, plant, and fungal pests that interfere with crop growth and spread disease. Elimination of these chemicals would likely cause famine and disease in many regions of the world. Thus, we must admit that these chemicals currently have a benefit to human health. At the same time we also must acknowledge that many, if not most, of these chemicals have the potential to exert deleterious effects on human health and the ecosphere. Judicious use of these chemicals is essential and steps should be taken to use the safest and lowest amounts possible, including the use of integrated pest management strategies. Currently, the relationship with environmental health scientists and companies that manufacture and utilize these chemicals is mainly adversarial. Some chemical companies have focused more on their own corporate health than that of the general public, but with the proper motivation these companies could develop practical and economically sound solutions to reduce the impact of their products on the environment. The development of the field of green chemistry has been very beneficial to the chemical industry, but has had little impact on pesticide chemistry. Implementation of green chemistry approaches could help maintain control of pests but help reduce the negative impact on the environment. Synthesis of safer compounds with fewer toxic by-products could benefit all involved parties. Partnerships between chemical companies and environmental health scientists could accelerate this process. Indeed, the regulatory agencies across the world would be wise to implement strong incentives to companies to develop green chemistry approaches to pest control.

4.7 OBSTACLES AND OPPORTUNITIES

Environmental health sciences is an umbrella discipline that is composed of multiple subdisciplines. Within the field communication among the population, exposure, and mechanistic scientists could be improved. Interaction among exposure scientists, toxicologists, and epidemiologists does not occur as often as one would hope. Enhanced collaboration among these areas is important for the health of the field, but it is especially important within the context of the exposome. The arbitrary boundaries that drive the reductionistic study of environmental factors in disease are eliminated, and their removal is likely responsible for some of the consternation from many scientists who work in the specific subdisciplines. Scientists have a tendency to be territorial and defend their scientific turf. While it is important to maintain the integrity of various scientific disciplines, science is moving and evolving at a rapid pace. Those poised to capitalize on the newest innovations within the context of their own discipline are likely to be in a stronger position in the future. Eventually, the defenders of scientific turf may look back to see that the grass is dying.

The broad field of environmental health sciences must be the driver of any exposome initiative and given the critical need to reach out to numerous outside disciplines, it is essential to have a coordinated structure to entice these other fields to examine the problems surrounding the exposome. Currently, the exposome is being discussed within the various subdisciplines, but there has been little coordination among them. This is an important obstacle to address. Indeed, for the outline provided in Chapter 7 to succeed, it must be driven by a unified environmental health science community.

4.8 DISCUSSION QUESTIONS

Why is it important to address the mechanistic basis for associations between given exposures and health or disease outcomes?

Does the exposome add value to the field of environmental health sciences or is it a distraction?

What is outside that sphere of environmental health sciences? Biological, physical, and chemical influences on health? What about behavioral, social, etc.?

FURTHER READING

Brunekreef B. Exposure science, the exposome, and public health. Environ. Mol. Mutagen. 2013;54(7):596–8.

Davis M, Boekelheide K, Boverhof DR, Eichenbaum G, Hartung T, Holsapple MP, et al. The new revolution in toxicology: the good, the bad, and the ugly. Ann. N. Y. Acad. Sci. 2013;1278:11–24.

Lioy P. Exposure science: a need to focus on conducting scientific studies, rather than debating its concepts. J. Expo. Sci. Environ. Epidemiol. 2013;23:455–6.

Rappaport SM, Smith MT. Epidemiology: environment and disease risks. Science 2013;330 (6003):460–1.

CHAPTER 5

Managing and Integrating Exposome Data: Maps, Models, Computation, and Systems Biology

5.1 MAPS, MODELS, AND TECHNOLOGY

Maps and models are abstract representations of reality. A map of a particular country is a representation of a country's physical features. The map provides depictions of geography, railways, highways, population density, and many other quantities. The map is not the country, but it provides incredibly useful information to the user, and that information can be used to navigate the country. In an analogous manner, models of biological pathways are mere representations of complicated systems. A model of the Kreb's cycle tells the user how the various constituents are modified as they move along the biochemical highway. It provides information about enzymes, cofactors, substrates, and products. The model can help the scientist navigate the biological landscape. Thus, from a thematic standpoint, maps and models are similar in their ability to help make the complex understandable, and both will be covered in this chapter. While many of the topics discussed may seem technologically daunting and potentially far-fetched, the reader is to be reminded of the extraordinary pace of technological advancement.

The year 1984 provides an example of the rapid advancement that has occurred over the past few decades. For the more seasoned scientist, *1984* conjures up images of a dystopian society from George Orwell's novel, a fate we have managed to avoid. For the mid-level scientist, it may jar memories of an Orwellian television commercial that appeared that same year (if you were born after 1984, you should probably read the book *and* watch the commercial). The television advertisement showed throngs of people with shaved heads and drab clothing silently watching a video screen of propaganda (a representation of a scene from Orwell's book). In stark contrast to the drab scenery, a young and brightly dressed woman runs past guards and throws a sledgehammer through the screen in the hopes of waking the masses.

The message was that Apple computer was not going to let 1984 be like Orwell's *1984*. It was the audacious introduction of the Mac. In retrospect, the audacity may have been understated. Coincidentally, 1984 was the same year when the first cellular telephone became commercially available. Weighing in at a whopping 2 lb and at a cost of $4,000 (nearly $9,000 when adjusted for inflation), the Motorola DynaTAC8000X was a far cry from what we have now. Referred to as the brick (its approximate size), the DynaTAC sported 30 min of talk time with 8 h of standby time. After steady and incremental improvement in cellular phone technology, the iPhone was introduced in 2007 and transformed the industry. No longer was the cellular phone a mere communication tool. It was a portable data management system. The technological power contained within our smartphones is staggering when one thinks back to what was available 30 years ago (at that same time I was learning to use computers on a Commodore Vic20 with 64 kB of memory and Tandy TRS-80 with a mere 4 kB of memory). While checking one's Facebook status while walking to class may not change the world, the potential to track one's movements along with a host of personal environmental variables (temperature, humidity, particulate matter, various gases, radiation, luminosity) and biological endpoints (body temperature, heart rate, sweat composition) could revolutionize spatial exposure assessment. All of these technologies are essentially possible today, and every person could eventually become a sentinel for dozens of key environmental indicators with their data uploaded and mapped in real time. Just imagine where the technology will be 30 years from now. Things that seem impossible now may well be common within our lifetimes. Many of our current intellectual restraints regarding the computational and analytical power that will be needed for the exposome are self-imposed. We should prepare ourselves for revolutionary technological advances … the exposome-tricorder perhaps?

5.2 MAPPING

The ability to geocode patient and exposure data provides an extraordinary opportunity to generate interactive maps of environmental exposures. Mapping data-rich exposome information will be beneficial to public health professionals and researchers. This information could also be very useful from a regulatory standpoint. By providing specific information on where particular exposures occur and their proximity to the suspected source(s), such approaches greatly expand our understanding of the dynamic nature of exposures. These types of approaches also begin to capture information about the impact of weather patterns on actual exposure. A person may exercise at a park one mile from a factory, but the exposures could fluctuate dramatically based upon their level of exertion and whether or not it has recently rained. However, all of this information could be readily detected with a really smart smartphone (geniusphones?).

The utility of mapping chemical use and exposures is nothing new. The field of remote sensing is grounded in geospatial analysis. Numerous organizations have been mapping exposures for decades. For example, the US Geological Survey has generated detailed maps of pesticide use over time. It is important to note that in this example, these are use maps, not exposure maps. Data for hundreds of pesticides are available over many years (Figure 5.1).

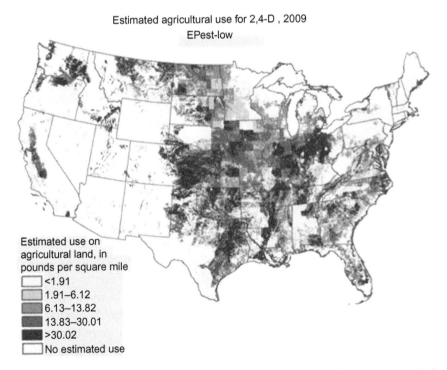

Figure 5.1 Sample map from the US Geological Survey Pesticide National Synthesis Project. As an example, the distribution of the herbicide 2,4-D is shown for the year 2009. With permission from Thelin and Stone (2013).

The US Geological Survey data come from the Environmental Protection Agency registry database. The data are provided at a county level and thus has relatively low resolution. At a country level, the data are visually impressive, but one quickly realizes that these depictions of seemingly catastrophic exposures do not necessarily translate into human exposure or alterations in human health. The maps certainly tell us about the use of this class of chemicals and reveal where most of our agricultural activity occurs, but we already knew this. While these maps are visually appealing, their resolution may be insufficient to draw meaningful conclusions regarding the exposome. It is likely that more detailed maps will be necessary. Dr. Beate Ritz at the University of California, Los Angeles, has generated and validated more extensive maps for the California Central Valley. Her data include distance between the actual pesticide application, based on data from the California Environmental Protection Agency (CEPA), and workplace and residence with detailed geocoding. Importantly, the group also verified the utility of their models by measuring the levels of selected pesticides in the blood of hundreds of workers in those areas. The results showed excellent concordance with the predictive models. However, these data are based upon the reported purchase (and the presumable use) of currently available pesticides and not the actual levels in the water, soil, and air. Banned compounds would not be reported in the CEPA data, but such compounds were identified in the biological samples, emphasizing the importance of including banned, yet persistent, chemicals in our analyses. Not only are they indicative of historic exposures, they may be driving adverse health effects. Analysis of current pesticide use may just be a proxy for past exposures to more toxic chemicals.

What really matters, though, is how much of these pesticides are getting into our bodies. This is where it becomes necessary to measure specific chemicals in human samples. There is likely to be a significant degree of concordance, as shown by the Ritz group, but we cannot make that assumption. We must conduct the experiments on a large scale and measure a wide array of chemicals in the human body. The exposome should certainly build upon these existing data sets when possible, overlapping them with census-type data for population, education, socioeconomic factors, and disease incidence, but it is critical to weave this information with quantitative assessment of exposure from humans.

The importance of maps in the context of the exposome has not gone unnoticed. Dr. Paul Juarez at the University of Tennessee has been using maps to study environmental health disparities. With a policy-based focus, his group is focusing on the data that are currently available to better understand disparities in health. His project entitled, "Research Center on Health Disparities, Equity, and the Exposome," has amassed 30 years of data on environmental factors, such as air and water quality, emissions, toxic waste sites, and more. It also includes data on the built environment, housing, social factors, and regulatory factors (zoning, state bans, etc.). His work has used a map-heavy approach focused on the population level to better inform public health; however, there is no biological data that can address the complex exposures affecting health (Figure 5.2).

There are many other potential uses of maps. The traffic flow data on your smartphone comes from information from thousands of smart-phones moving along the highway. This technology can provide data

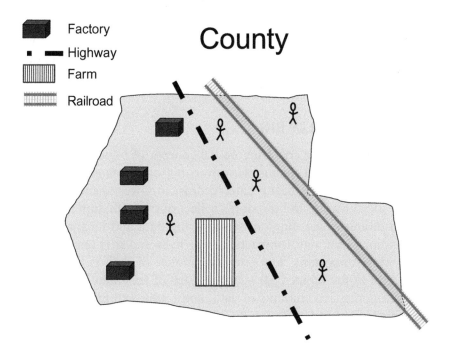

Figure 5.2 Simplistic depiction of a map of exposures. One's proximity to factories, farms, and highways will impact one's exposures. For many years, scientists have used the location of residences and workplaces to estimate exposures. The technological advances and availability of GPS-enabled smartphones allow real-time location. This should greatly increase the resolution of exposure-related mapping.

on mobility and health. A truck driver's phone is going to log 10 h of moving at 50 mph, which is too fast for exercise and suggests the user is actually sedentary. A jogger may log 30 min/day at 6 mph. An office worker could spend 10 h a day at 0 mph (at desk). A construction worker may spend 10 h ranging from 0.1 to 1 mph as he or she moves around the work site. Each of these activities (driving, exercising, working) affects health in different ways, and the generation and collection of these data are relatively simple. It should be possible to automatically determine the type of activity without any input from the user (a bin of 10 mph activity, while possible on foot, would be excluded as physical activity if it was flanked by bins of 50 mph activity, as it is likely a determinant of traffic). We may not want this type of information being sent to our car insurance company as it could also be used to determine reckless driving and speeding (vector patterns should be able to discern driving patterns), but from a health perspective it could be informative. Similarly, a health insurance company or employer could be interested in visits to fast food restaurants or bars. It would just take the merging of some of the technologies currently used on GPS and mapping programs on smartphones. Regardless of how these spatial data are used, it is clear that they will generate massive amounts of data and some of the issues surrounding that will be discussed below.

5.3 BIG DATA, REALLY BIG

Years ago most people could ignore the concept of big data, leaving it to the statisticians and quants. However, big data sets are infiltrating our everyday life; from health care records, to government retirement plans, to social media, to how stores collect our information and anticipate our desires (marketing on steroids). Big data and its analysis and interpretation have major impacts on our lives and it is only going to become more important. The growing amount of data is becoming a critical issue in the biomedical sciences. Each of the -omic technologies is generating massive amounts of data, and traditional means of analysis are insufficient to handle it.

In addition to the -omic activities, the aforementioned mapping approaches also generate incredibly large data sets. The ever-expanding data sets must be integrated into organizational frameworks. Computational and systems biology approaches will be essential for such an endeavor. Many organizations recognize these concerns and are working on solutions. The Europe Commission is funding a Big Data initiative (big-project.edu) called the Big Data Public Private Forum. The US NIH is also establishing Centers of Excellence under the Big Data to Knowledge (BD2K) program. Some of the approaches mentioned in Chapter 3 are already working with terabyte to petabyte data sets. As we merge the data from the various -omics, it will certainly push us beyond the exabyte into the head-spinning zettabyte and yottabyte range (Figure 5.3).

Number (bytes)	Abbreviation	Name
1000	KB	Kilobyte
1000^2 or 1,000,000	MB	Megabyte
1000^3 or 1,000,000,000	GB	Gigabyte
1000^4 or 1,000,000,000,000	TB	Terabyte
1000^5 or 1,000,000,000,000,000	PB	Petabyte
1000^6 or 1,000,000,000,000,000,000	EB	Exabyte
1000^7 or 1,000,000,000,000,000,000,000	ZB	Zettabyte
1000^8 or 1,000,000,000,000,000,000,000,000	YB	Yottabyte

Figure 5.3 The impending yottabyte. Current experiments that evaluate environmental exposures are already generating terabytes of data. When these data sets are combined with those from genomic, proteomic, and epigenomic studies, we will undoubtedly be looking at petabytes and exabytes of data, and it will only grow.

While the challenges facing the exposome are not yet on the proverbial map of these big data initiatives, we can certainly take advantages of the tools and approaches that are spurred by these projects. New exposome data must be amenable with the systems and approaches that are being developed for the genomic and proteomic initiatives. Indeed, a major goal for exposome enthusiasts should be to develop partnerships with these big data centers that are emerging for other fields.

We will also need to come to grips with computers running larger parts of our lives. Hal in 2001: A Space Odyssey, or Big Blue playing chess or winning on Jeopardy may not seem threatening, but when such systems are diagnosing our illnesses, deciding our medical care, or determining our retirement allocation we may start paying more attention. I believe it is incumbent upon the users of these systems to demonstrate how they solve less complex problems in ways that are logical and scalable. The ability of these systems to make obvious decisions in a way that is verifiable should be demonstrated to the people that will be affected by it. The decisions will be too important to ask the public to trust the organizations that are using that information to influence our lives. The decisions made by these systems for complex problems will continue to be evaluated through peer review and expert panels, but providing information on the systems to the ultimate affected user (e.g., taxpayers) should be required.

One of the author's favorite scientific essays is Chaos in the Brickyard by E.M. Forscher. In the parable, builders are the metaphorical scientists. Initially, the builders crafted their own bricks to their exact specifications. Later, they employed junior scientists to serve as brick makers to expedite their work. With proper supervision, high-quality bricks continued to be made. At some point, the craft of brick making became as valued as the craft of building itself, and thus a new industry developed for the task of making bricks. Bricks were churned out at an ever-increasing pace without effort being put into the actual construction of buildings. There were too many bricks and too few buildings. Eventually, the ground was covered in bricks making it impossible to construct new buildings. The essay was written over 50 years ago. Watson and Crick had just been awarded the Nobel Prize. The writer could not have imagined the current state of the brickyard. Given that so much information relevant to the exposome is already present in the pile of bricks, will it be possible to find a clear spot to establish a foundation? It is not clear how a space can be cleared to start building the exposome, but it seems obvious that the blueprints, foundation, and framing must involve computational approaches. There are just too many bricks.

5.4 I AM SO SMART, S-M-R-T

Humans are incorrigibly inconsistent in making summary judgments of complex information.
Daniel Kahneman in "Thinking Fast and Slow"

A Nobel Laureate in Economics may seem like an odd source for insight on the exposome, but Dr. Kahneman's statement highlights the need for computational systems in evaluation of the exposome. The exposome is complexity at its finest. Humans are notoriously poor at evaluating large amounts of data and making accurate and consistent decisions. Our brains are wired to simplify complex problems into apparently similar, but more manageable problems, often referred to as heuristics. What we refer to as intuition is actually complex associative processes occurring subconsciously in our brain. It is an impressive level of computation, but it is also fraught with bias and irrational decision-making. We are lulled into a false sense of security and think that we are making wise decisions and sound predictions, but in fact, we fail miserably. Just look at findings from studies of prognosticators across disciplines. Even professionals such as stock traders, political pundits, and business experts rarely perform above statistical chance in the long term (and arguably get paid better than their performance warrants). Many of these complex systems tend to defy predictions anyway, but somehow we continually believe we can do it with aplomb. Much to our chagrin, cold and robotic algorithms consistently outperform human intuition in medicine and other important decision-making processes. When dealing with the exposome, human intuition will not be sufficient. We must come to grips with our shortcomings and be ready to place our cautious trust in machines.

5.5 MATHEMATICAL MODELING

It is likely that terabytes of exposome-related data will accumulate before we have had the chance to build a system in which to store and organize it. What is even more intriguing is that so much exposome-related data are being collected even before the overarching and organizing concept of the exposome has been fully embraced. It is not as if there is an alternative concept to organize the exposure-related data. I argue that scientists must take a step back and work on constructing the foundation for such a project. It may be necessary to construct small buildings of focused relationships while we figure out what the neighborhood block will look like.

Given the past, current, and likely future states of computational and systems biology (recall cell phones were not in use until 30 years ago), I would argue that the initial plan should be unrestrained: a truly idealistic and optimally functioning model. For example, a model that can quickly assimilate data collected in real-time from the individual, such as activity, behavior, food intake, and personal environmental sampling, with data from somewhat invasive analysis of biological samples for metabolomic, (epi)genomic, transcriptomic, and targeted chemical analysis. These data could then be entered into the established exposome model to provide a personalized risk profile for thousands of diseases, along with a plan of how to improve the exposome profile.

Such a deliverable may take many years to arrive, but the goal will start help identifying the computational technology, approaches, etc., that will need to be developed. This list of desirable tools could serve as a driver for innovation within the field, or could ready the field for rapid adoption into the exposome model once it becomes available. The mega-model may take decades to complete, but within its framework, there will be great opportunities to develop some of the small models that can become part of the larger system and provide the organization system to contextualize new findings.

It is highly likely that the modeling of the exposome is going to be a multinodal design, but it should be designed in a manner that allows integrations of the various parts. For example, one model may focus on the physicochemical interactions of chemicals in our immediate environment, while another may focus on the transport of those chemicals into the body and the subsequent kinetics of the mixture throughout the body. Others may focus on biological effects at an organ systems level, with each organ or its major functions being modeled. Many of these functions result in alterations in metabolites that reach the bloodstream, for example, reaction products from liver metabolism, chemicals reclaimed or excreted from the kidney, or altered glucose from pancreatic activity. Thus, an exposome model that incorporates the metabolic features found in the blood could capture many of the biological functions occurring within our bodies.

Computational approaches have been used to develop pharmacophores and toxicophores, and this is how some of the Tox21 data are being modeled and analyzed. The idea is to use mechanistic toxicology and structural biology to figure out the key properties and then to build computational models that can predict the likelihood of unknown chemicals affecting the target. These computational models can then be used to study the hundreds of chemicals present in the human body to determine the potential additive and synergistic effects. For example, we know that organophosphate insecticides inhibit acetylcholinesterase. We further know that dozens of these products are on the market. If a person has 24 different organophosphates in their bloodstream, should we not evaluate them as a group? Such approaches have been used to some degree at the regulatory level, but it will require more sophistication to make connections between these molecular interactions and disease states. Addressing these problems will require the use of computational models.

5.6 MODELS

Developing a model is a complicated process. The modeling process has several major steps. The first is to identify the goals of the project, including the scope. While much of the discussion of the exposome has been boundless, limits will need to be identified when constructing models. What is it we want to model? Are we focusing on exposures first and addressing the various -omic projects later? Or do we start with the wealth of genomic, transcriptomic, and proteomic data that is already available? These are not simple questions, but framing the question is a critical step and one that should be carefully decided. Once the goals are identified, it will be necessary to identify the currently available data and what we already know about how the environment impacts health. The next step is selecting the type of model. This decision will be based on a series of issues surrounding the type of data that are presently available and the type of data that is expected to become available. Since time is an important variable, whether it be diurnal variations, day to day alterations, or yearly to decade changes, the model will need to be dynamic. Some components may be mechanistic, for example, in a biochemical pathway, but some may correlative, as in how an alteration in a biological pathway is related to a health outcome without knowing why. A regression-based model can make reliable predictions, but offer no mechanistic insight. While a multiscale model that could model molecular pathways and complex human behavior would be nice, the likelihood of such an approach working is very low. Thus, it may be necessary to revert to our reductionistic ways and develop more feasible models that examine one specific type of exposure or endpoint in the hopes of future integration. It is rather difficult to assess the quality of the data that are currently available because we do not know what it is we are trying to do. Ideally, once the objectives and scope are identified then data standards can be developed to determine which data would be suitable for analysis.

The actual selection of the model(s) is critical. First we must determine the essence of the complex exposome. For example, we can state that external factors alter human biology. This is extremely broad, but it establishes the relationship. We need a way of modeling the external factors and a way of modeling human biology. Our model could benefit from having correlative and explanatory components. If an increase in one particular external factor decreases function of a certain biological system, we do not necessarily need to know why. If we did, it could help identify similar contributors and allow us to determine what factors are influencing the accuracy of the predictions. Given that we do not have all of the necessary information available at the outset, a deterministic model will not be appropriate. A probabilistic or stochastic model is better at dealing with randomness and variability, and those are certainly going to be components of the exposome. There are a large number of variables, signals, and parameters that will need to be defined. Once a model is sketched out and populated with various types of equations it needs to undergo thorough diagnosis to determine if it is internally consistent, sensitive to changes in input, and stable under such challenges. It must be determined that the model is providing data that is reasonable at a qualitative and quantitative level. The next step involves the actual use of the model, which starts with validation using experimental data and then moves to untested scenarios. Based on those experimental results, the model can be manipulated and optimized. Thus, it is an iterative process with multiple checks and balances. Ultimately, the goal is to have a model that provides accurate predictions of outcomes based on novel inputs. For example, does a particular combination of chemicals and stressors increase or decrease the likelihood of contracting a particular disease? This process is not for the faint of heart. As Dr. Eberhard Voit, when referring to the exposome, has stated, "This is not rocket science, this is harder than rocket science. Space exploration is based on linear models, the exposome is based on non-linear models, which are much more complicated." Harder than rocket science. Great.

5.7 PREDICTING PREDICTIONS

Predicting outcomes is a commonly stated goal of many health-related projects. We want to predict who will develop diabetes. We want to predict who will have a heart attack. We want to predict who will develop cancer. We certainly know that having an increased number of risk factors increases the likelihood of a person getting a disease, but in reality we are not that good at such predictions. Are we being unrealistic in thinking we can predict our exposome-regulated destiny with any level of precision? One option would be to stop trying. Rather than focus on predicting outcomes that occur in the future, perhaps we should try to measure an individual's ability to respond to those exposome-encompassed external forces in the here and now. Which is harder to do, predict who will get cancer in the next 20 years or who will respond more favorably to an environmental insult in the next hour? It is very difficult for a cardiologist to predict who will die of a heart attack in the next decade, but she or he can perform a stress test on a patient and know immediately how the patient responds to the challenge of a bout of exercise (external stress is the work performed) and assess the patient's robustness or fragility with a high degree of precision. We can try to predict who will perform poorly in such a test, but we do not stop there. We actually test them. We do not rely on predictions. In the context of the exposome, we want to know how people respond to their past and current environmental exposures. We want to know what the cumulative biological response is and use that as a measure of fragility.

5.8 OBSTACLES AND OPPORTUNITIES

One of the great opportunities for the exposome is the possibility of generating maps based upon exposure data and biological data, but this level of integration will require sophisticated analytical tools. Groups that study weather patterns and atmospheric modeling are dealing with similarly complex data, and collaborations with these groups represent an excellent opportunity. Over the past few decades, the ability to accurately assess weather patterns has grown tremendously. While weather patterns may become an important component of the exposome, it may be the techniques borrowed from that field that may prove to be the most useful. For example, the US Department of Energy's Oak Ridge National Laboratories and the National Aeronautics and Space Administration have extremely powerful computer systems to process spatially related data. Similar to Human Genome Project, government laboratories across the world will need to be brought into the fold.

Drs. Germaine Buck-Louis and Rajeshwari Sundaram have written about the importance of exposome research in a recent commentary in Statistics in Medicine. They introduce the exposome paradigm to the statistically-based audience and make the case for the need of the exposome to address the major gap in our knowledge of environmental impacts on health. The authors argue that it is time to move forward with such an initiative in that the environment is being left out of much of biomedical research. The authors also suggest that it is important for the exposome to be inclusive of all exposures irrespective of source, timing, and duration, and that external and internal environments must be measured. The authors suggest that the exposome is a critical missing element in the developing mega-database on human health (what comes after yottabyte?).

If systems and computational biology are going to become part of the exposome foundation, the field of environmental health sciences will need to start providing training in these areas. While it will continue to be necessary to rely on specialists in these areas to some degree, the investigators interested in the exposome will need to be familiar with the language, tools, and strategies used by the quantitative specialists. This may be one of the more significant obstacles for the exposome in that environmental health scientists are already embracing the -omic type technologies, spatial statistics, and advanced epidemiological analyses, but systems and computational biology do not seem to be members of the club.

5.9 DISCUSSION QUESTIONS

If detailed GIS-level information on environmental exposures was freely available would it impact where you choose to live? Would it affect property values?

How comfortable are you with relinquishing decisions about your health to a computer? Are their situations in which you would refuse to accept such input?

Devise a way to evaluate the utility of a computational approach to the exposome? What aspects are necessary and how would you test for them?

FURTHER READING

Buck Lewis GM, Sundaram R. Exposome: time for transformative research. Stat. Med. 2012;31:2569−75.

Chen R, Snyder M. Systems biology: personalized medicine for the future? Curr. Opin. Pharmacol. 2012;12(5):623−8.

<http://communitymappingforhealthequity.org/public-health-exposome-data/>.

Thelin, G.P., and Stone, W.W., 2013, Estimation of annual agricultural pesticide use for counties of the conterminous United States, 1992−2009: US Geological Survey Scientific Investigations Report 2013−5009, 54 p.

Voit EO, Qi Z, Miller GW. Steps of modeling complex biological systems. Pharmacopsychiatry 2008;41:578−84.

Voit EO. A First Course in Systems Biology. New York, NY: Garland Science, Taylor and Francis; 2003.

CHAPTER 6

The Practical Exposome: Education at the University and Community Level

6.1 IMMEDIATE PRACTICAL UTILITY

As noted in Chapter 5, the challenges facing data analysis for the exposome are daunting, some may say impossible. I ask the reader to set the thoughts of impossibility aside for a moment. There is one aspect of the exposome that is not complex or fraught with challenges and scientific danger. It is my view that the exposome is a superb educational vehicle to promote the importance of the environment in health and disease. It is not necessary to have hammered out all of the details. The exposome is an accessible mental model that can be used by students and citizens alike. Each day we are bombarded by a dizzying amount of exposures and influences from our environment. We are aware of many of these exposures through our senses and intuition, and also through information coming from the media and our personal network.

Recall that human intuition fails miserably at dealing with complex data. Most people cannot keep track of the various controllable environmental factors they face every day. Keeping track of a diet is one thing, but remembering not to exercise during rush hour or near crowded highways, to avoid the stress of overscheduling, to avoid mixing certain household chemicals, to thoroughly wash fruits and vegetables, and to take any necessary prescriptions, can be exhausting. Most people throw up their hands in defeat and just focus on what they perceive to be the most important. Unfortunately, their perceptions, which are shaped by the media, advertisers, and well-meaning friends and family members, are usually shaky from a scientific perspective. The exposome, especially charted as the modifiable version shown below, is an excellent way of organizing these multiple influences in a way that brings clarity to the morass of information. The development of an expanded, personalized, and easy-to-understand version of this could become an interactive tool to be used with one's physician, as a guidepost for home eating habits and as a blueprint for overall health.

6.2 THE MODIFIABLE EXPOSOME

The components of the exposome display varying degrees of impact on one's health. Cigarette smoking is much more deleterious than the occasional hot dog. Some components are very much under our control (smoking, physical activity), while some are influenced more at a population or societal level (air pollution, health care access). Figure 6.1 illustrates the relative contribution based on these two scales. The y-axis shows the relative degrees of influence, while the x-axis displays the level of control from individual to population. Items in the upper half of the diagram have major influences on health and we should be taking steps to minimize their negative impact or to enhance their benefits. Items on the bottom can certainly influence health, they just take a lower priority. The items on the left are under the control of the individual or the community to some degree, while those on the right are more on a population or societal level.

The top right quadrant contains the major issues that must involve governments and health care systems. Improving health care access, reducing air pollution, improving water supplies, all require some intervention on a large scale that involves populations, governments,

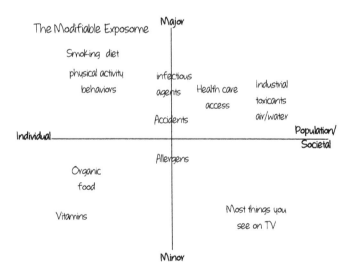

Figure 6.1 The modifiable exposome. The exposome provides a useful mental model to organize the multiple influences on our health. By using an intentionally broad definition of environment, nearly all of the external factors that influence one's health are included in the exposome. It is then possible to plot these various factors along axes that include the importance to health (minor to major) and the ability to control said factors on the individual, community, or societal/population levels.

and regulations. Of course the individual has a role in voicing concerns to their elected officials, but the ultimate solutions come from the larger organizations. Climate change was left out of this graph. This was not intended to minimize its importance, but because the time frame of the graph is relatively short. The idea is to explore how people can modify their exposome in a positive way on the scale of days, months, and eventually years. This may be temporally naïve, but it allows the reader to focus on the more immediate future from a personal health standpoint. The proper care and maintenance of our planet is crucial, but it is beyond the scope of this introductory text.

Eating organic foods is listed in the bottom left. The author is likely to catch some flack for this. Overall, the judicious use of chemicals increases crop production and keeps disease in check. From a global perspective, organic produce is a luxury. Certainly if one can afford it, it is best to minimize one's exposures to pesticides and other agricultural chemicals. But eating organic is not as important as eating a healthy diet, and for that reason it is viewed as minor. The benefits of eating an organic diet may be as much about the quality of the food itself and less about the lack of pesticide residue. A move toward lower pesticide use across the entire agricultural sector would be of greater benefit than the expansion of the organic industry (although the latter may be helping drive the former).

Accidents straddle the line between individual and population because there are individual steps (seatbelts, protective eyewear, safe practices, common sense) and societal steps (OSHA regulations, workplace practices) that can be taken. Allergens are listed below the x-axis because for the majority of people allergens have a minor effect. Of course for individuals with severe allergies, the allergens may pose a much greater effect, even being one of the most significant drivers of health. Obviously, if such a graph was personalized and accounted for the individual susceptibilities and vulnerabilities, the layout could look much different.

It is important to remember that each axis is a continuum. We have some level of control over our exposure to toxicants in the air, for example, maintaining vehicles, not exercising near highways, and using water purifiers depending on water source. However, the major drivers (pun intended) are at a higher level. The number of vehicles on the road, exhaust regulations, and industrial pollution controls must be addressed at the regulatory and societal level. While cigarette smoking is on the left side, we know that community and societal issues are also at play. From a societal standpoint, cigarettes should not be available to children and teenagers. We know that the sooner people start smoking, the more likely they are to become addicted. While raising the drinking age to 21 in the US was a slow process and required significant governmental intervention, it did occur and has had a positive impact on the rates of accidents and injury. Had the US tobacco settlement included a similar raising of the legal smoking age to 21 (phased in over several years), it would have done more to improve the health of the US than all of the billions of dollars that were poured into state coffers (even with a predictable prohibition-like black market).

The "most things on TV" is only partially tongue in cheek. The vast majority of news stories related to health are sensationalist. They focus on obscure, but newsworthy exposures. Decades ago I recall seeing a television news piece on the Marlboro man that included a critical assessment of the tobacco industry (this was when information regarding the cover-up of adverse effects by the industry was becoming public). Today, it is rare to find anything on television that is remotely educational concerning environmental factors in health.

Infectious agents are listed very close to the top of the graph. Of the items in our environment, infectious agents have the potential to have the biggest immediate impact on health. While many factors involved in the spread and contraction of disease are outside our control, the individual can be sure to obtain recommended vaccines, including seasonal flu vaccines, avoid contact with sick people, and follow appropriate hygiene practices. But governments and health care systems must make these vaccines available and promote the appropriate hygiene practices. Communities must work to encourage these important activities, and professionals must continually work to counter the pseudo-scientific claims that pervade the public thinking.

The exposome model is also an excellent way to address health disparities. Variations in the factors that contribute to the exposome are likely to be greater than the genetic factors when one looks at various populations within a particular region. Poorer areas tend to be in closer proximity to hazardous waste sites and farther from parks and health care facilities. As noted in Chapter 5, Dr. Juarez and his research group are using an exposome-based approach to evaluate such disparities and how they may impact health. Layering rich exposome-type data on top of census and other socioeconomic data can help reveal trends among various exposures. The addition of comprehensive biological data could bolster these analyses, but it will be necessary to maintain a strong geospatial component to exposome data collection to facilitate the layering of these data.

6.3 BEHAVIORS, MANDATES, AND NUDGES

The concept of nudge, which garnered widespread attention with the publication of Nudge: Improving Decisions about Health, Wealth, and Happiness by Richard Thaler and Cass Sustein, is to gently guide people into positive behaviors by making it easier to make the right decision. The decisions that people make regarding their environment are based on a combination of factors. First, the individual can know what is best for him or her and pursue that particular action. Second, the individual can know what is best for him or her and chose not to pursue that particular action. Third, the individual can be ignorant of what is best for them and they may or may not pursue the appropriate activity. For each of the dozens of choices to be made in a single day, there will be various combinations of the aforementioned decisions. Often the poor choices we make in maintaining a healthy exposome are based upon our own laziness. We may well know what is best, but because such decisions can require more effort and energy, we choose not to use them. The conservation of energy may appear to be a positive adaptation from a bioenergetic standpoint, but it leads to small negative consequences that can build up over time to weaken the individual. Thus, there is an advantage from a policy, regulatory, and organizational perspective to make good choices the default option. We see this sort of strategy used in many life settings. For example, the opt-out set up of a company 401k makes the default to save money. One must consciously choose not to participate in a retirement plan. When participation in such a program is opt-in, the participation rates are much lower.

Antismoking policies are designed in a similar manner. The default behavior in settings with such restrictions is not smoking. While the basis of such programs is often framed in the manner of reducing second-hand smoke, which is a good thing, it also forces behavioral changes in the smoker. With the majority of the world population being nonsmokers, antismoking policies are well accepted by the general public. The general population notices that there is less smoke in their presence and there is no downside. Things get trickier when we try to encourage behaviors that include the majority of people.

A recent example of this was the school lunch initiative in the US. It made perfect sense. Our children need to eat healthier food. If only healthy choices are made available, the children will have no choice but to eat healthier food (in general, the healthier choices have lower levels of preservatives and higher levels of nutrients). The US Department of Agriculture (USDA) has control over the school lunch program so instituting broad changes was easy. The result was a major pushback by children and school systems due to the swift and dramatic implementation. Had they put the frog in cold water before turning on the stove, the changes could have been slowly implemented without the students noticing. The concept of nudge in the hands of a politician or a federal government can quickly turn into the concept of shove. This may explain the dramatic pushback to Mayor Bloomberg's attempt to reduce the general public's consumption of sweetened beverages in New York City. While reduction of sugar intake is good for health, creating rules and laws that make swift and dramatic changes in behavior do not go over well. Sustein extolls his accomplishments regarding the development of easy-to-understand government policies during his term in a subsequent book *Simpler*, although he did not acknowledge that the Affordable Care Act, which was introduced while he was still overseeing the release of regulations, was an example of precisely the opposite of what he claimed to be doing.

6.4 ACADEMIC EDUCATION: THE EXPOSOME IN THE IVORY TOWER

Within the field of public health, most schools have very little basic science in their curriculum. Basic concepts in biochemistry and physiology are often overlooked because it is not clear why students need this information. A course that examines the balance of genetic and environmental contributors to disease can provide an excellent introduction to many of the important scientific concepts in public health that make it easier to understand the various drivers of health. The exposome provides an organizational structure to understand the complex exposures that impact health, exposes the student to cutting-edge science from multiple disciplines, and forces them to think in a holistic manner. It reminds them of the rapidity of scientific advancement and serves as a reminder to keep up with the literature.

While medical students learn about human physiology *ad nauseum*, they hear little about air pollution and atmospheric chemistry, environmental and occupational health, and physical activity. For the most part, the medical field has been dismissive of holistic medicine practitioners, such as homeopaths and chiropractors, based on the tenuous scientific foundations of their disciplines. But medicine misses the main point. Individuals in the general population yearn to be treated as a whole person, and these aforementioned practitioners give them what they want. The average person does not want to be looked at as a renal system, a digestive system, and a cardiovascular system. They know that certain activities in one area of their life can impact another. Sending a person to four different specialists pretty much assures that they no longer understand the causes and effects of their actions to their health. It is critical to distill the complicated medical *minutia* into structures the general medical consumer can understand and appreciate. Since the exposome encompasses all of the malleable components of our health, it can serve as an excellent rubric to understand health, and the behaviors and actions that improve health or contribute to disease.

When I first started using the exposome concept, I was fearful that the general public would bristle at such an esoteric scientific concept as the exposome, but I have found quite the opposite response. It has more been along the lines of "Well, duh, we know this, we have just been waiting for the scientists to figure it out." The vast majority of people want to understand the various factors that affect their health, but we need to work on our delivery of the information. With environmental health not being taught in most US medical schools, we cannot rely on the medical field to deliver this message.

An example of a misunderstanding of the external factors that influence our health is the use of nutritional supplements and vitamins. A story in the monthly magazine *the Atlantic* by Paul Offit heaped praise upon Linus Pauling, one of the greatest chemists of all time (another one of those double Nobel winners, although one was for Peace so he has nothing on Sanger, Curie, and Bardeen), but equally biting scorn for his promotion of high doses of vitamins. Current research suggests that such vitamin regimens likely increase the risk of death, but the general public has a view of "well if vitamins are good then more must be better" and the health practitioner does little to sway this way of thinking. Most people cannot fathom that supposedly healthy vitamins could be worse than exposure to low levels of pesticides (not saying it is, just that it could be). Swallowing gram quantities of vitamins, which obviously enter our body with ease, may well be worse than exposure to pesticide drift. An old health adage is that once there is adequate caloric and nutrient uptake, removal of factors has a greater effect on health outcomes than adding them. Removing a deleterious influence typically trumps adding a positive influence. To put it another way, eliminating bad habits is likely to have a greater benefit than adding good habits, and it is possible that ingestion of megadoses of vitamins may be one of those bad habits. It is the proper management of one's exposome that is key. People put so much literal and figurative stock into the vitamin industry, but much of this is due to misperception and aggressive marketing. In the context of the exposome, the use of vitamins may be an important part of one's health regimen, but if the nutrient level is high, it is more likely than not introducing a deleterious iatrogenic effect. By looking at all of the sources of chemical exposure in one's life, the individual exposures can be placed in the appropriate biological context. In fact, removal of excessive nutrients and pesticides would be the best outcome.

Another area in which the exposome can help frame health effects is the built environment. Urban planners know how to design a community that includes walking trails and parks. Architects know how to design buildings that have minimal impacts on the environment. They know how to utilize more environmentally friendly building materials. They know how to make use of mixed-use developments. It is unfortunate that governments do not do more to reward more intelligent community and housing developments. Leadership in Energy and Environmental Design (LEED) certification promises users that future cost savings will offset the increased building costs, but communities also benefit from this type of thoughtful construction. The private organization has been instrumental in getting more environmentally friendly building to occur, and more recently has introduced the concept of LEED communities, which encourage more expansive activities, such as redevelopment of sites in need of remediation from past toxic waste exposures.

6.5 INTRODUCING THE EXPOSOME AS A SINGLE LECTURE

For a single 1-h lecture, the exposome can be introduced as a novel approach to environmental health sciences. Beginning with the origin and definition, one can emphasize the importance of the environment by explaining the limitations of a purely genetic approach. The subdisciplines of environmental health sciences and their corresponding scope and expertise can be reviewed, followed by an introduction to how various investigators are approaching the complex exposures we face, including many of the -omic technologies. This leads into the challenges of large data sets and the need of systems and computational biology-based approaches to address the massive amount of data, ultimately converging on an improved understanding of how the environment impacts human health and disease.

6.6 THE EXPOSOME AS A UNIT WITHIN A COURSE

The exposome as a unit within an existing course on environmental health, toxicology, or epidemiology can be developed within the course context. Within a toxicology course, the exposome provides an excellent framework for toxicogenomics and high-throughput approaches. Within a course on environmental epidemiology, the exposome can be presented as a means of bringing higher resolution and more accurate environmental exposure data into research projects. For the exposure science course, the exposome introduces ways of integrating complex exposure data and explores the comparison of external and internal measures of exposures. In a general course on environmental health, the exposome provides a means of discussing how science is able to measure so much more than it could decades ago, and that this increased detection could dramatically enhance the field.

6.7 THE EXPOSOME AS THE BASIS FOR AN ENTIRE COURSE

Depending on the expertise at the institution, this course could be taught as a 1-, 2-, or 3-credit course. A 2-credit format allows an hour of lecture material each week followed by an hour of discussion of an assigned paper. At Emory University, we have been attempting to get the message out to as many students as possible. As such, we offer the course in a 2- and 3-credit format. The 2-credit format uses the hour lecture followed by an hour of discussion of an assigned reading.

Master's students and senior undergraduates participate in this. For doctoral students, we have them attend the same 2 credits, but we enlist their help in leading the small group discussions of the assigned readings. To provide more in-depth analysis of data in the field, the doctoral students also have a separate 1-h meeting where they critically examine an additional reading (although this could be on the same assigned reading but in much greater depth).

The concept of the exposome is pertinent to a wide variety of disciplines, and it has been our view that we want to use the exposome to teach about the importance of our environment to as many people as possible. Thus, a senior student outside the basic sciences may have a difficult time with a routine doctoral-level paper discussion, but they are comfortable with a discussion format that is more contextual, that is, not critically analyzing the experiments, but focusing on putting the findings within the context of the exposome (Figure 6.2).

The sample syllabus starts with an overview that would contain much of the content in Chapter 1 of this book. It sets the stage by

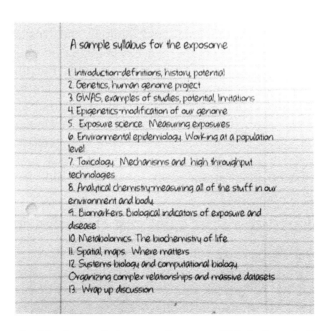

Figure 6.2 A sample syllabus for an academic course on the exposome. While the chapters in this book do not follow the exact order of the proposed syllabus, lectures can be supplemented with the text and the suggested reading material.

defining the concept and discussing the potential utility. The next lecture addresses the human genome. Depending on the backgrounds of the students, this may or may not be a review. Except for those focused on genetics and molecular biology, it is likely that the pace of this work leaves room for new knowledge. A lecture that explains the current state of the art in human genome sequencing is a very good foundation for the following lectures. The next lecture can go into how one can use genome-level data to study complex human diseases. The GWASs have been an important first step. Many have uncovered new genetic associations and others have confirmed previously identified linkages. The relatively new GWAS approach, which is based on SNPs, is already giving way to studies that utilize deep sequencing of specific gene regions. Having been exposed to a couple of lectures on genetics, the transition to epigenetics is smooth. It is important for the student to understand how the genome can be modified, but more importantly they should learn how readily environmental factors can do so. Once it is apparent that our genome can be manipulated by our environment, delving into how scientists assess those chemicals in our environment is a logical next step. From here, introducing approaches in environmental epidemiology is appropriate. For some students in public health, this may be too elementary, but with appropriate readings it is possible to provide a challenging lecture. Building upon the associations identified by the epidemiological methods, the next step is to address the exposures, biological plausibility, and mechanisms of action. Lectures on exposure science and analytical chemistry can be presented with specific information on how these samples are collected in the field and analyzed in the laboratory. Here is it very important to explain how mass spectrometry works and why various separation techniques (GC/LC) are needed for different types of samples. A lecture on biomarkers and how they can be used to track current disease states or past or current exposures helps make the connection between analytical chemistry and human health. Discussions can focus on what constitutes a useful biomarker and how such biomarkers can be used in medicine and health.

It is the opinion of the author that metabolomics may be one of the most important conceptual approaches to the exposome, as it is measuring the final common pathway of all exposures and biological reactions. Devoting an entire hour to this approach is certainly warranted. While a general lecture on the field of toxicology may not be appropriate, a lecture that explains the importance of demonstrating biological plausibility and an introduction to newer high-throughput technologies to measure the impact of hundreds of chemicals is an excellent fit in an exposome-based course. From there the course moves into more mathematical approaches. Spatial statistics, global positioning systems, and maps provide examples of how the very complex data can be organized. I must say that many students indicated that having the systems biology lecture at the beginning of the course would have made more sense. In essence, the students went through the semester wondering how it could be possible to deal with all of the complex data. For many it was a Eureka moment. The grandiose vision introduced in the first lecture did not seem as daunting with the possibility of systems biology (which was foreign to nearly all of the students). The intuition-exceeding processes and calculations had potential computational solutions. This lecture does not even have to be given from an exposome-based perspective. Just having a scientist explain how computational approaches can be used in biology is sufficient, the application to the exposome is obvious, and assigned readings and discussion can explore how the approaches can be applied to exposome-related problems. By the last class session, the students have been exposed to the possibilities and the problems. The instructors will have spent 12 weeks teaching the students about the exposome. The 13th week the tables turn. The wrap up discussion can start with a question as simple as this: if you have been asked to head up The Human Exposome project where do you start? The students are as up to date as most academic scientists. In fact most of the students are not constrained by traditional models of science and retain a level of idealism that facilitates this type of discussion. It is the upcoming generation of scientists that will be conducting the studies and translating them into public health. Their ideas are as valuable as those in the minds of scientists that learned science before the human genome was sequenced.

6.8 THE EXPOSOME OUTSIDE OF THE CLASSROOM

Academic investigators, well trained in the art of dissemination among their colleagues, often fall short when it comes to explaining their research to the general public, and that may be too generous. When one looks at recent stories regarding environmental factors in health, one sees more sensation than information. How valuable is it to dribble out information on environmental health exposures to the general public?

Given the current state of knowledge in the public, one could argue it is a terrible approach. The public will remember the sensational story about one type of chemical and then irrationally never use nonstick cookware (and then use oil on a traditional pan for decades). The public is fickle. A particular food is associated with a decreased risk of cancer and it becomes the trend *du jour*. Would not it be preferable to teach a mental construct that allows a person to frame a particular exposure in its proper context? Fresh fruits and vegetables are a key part to a healthy diet, and it does not matter if a study shows that broccoli or blueberries reduce cancer—you should be eating them. If a person learned to look at the current health report and weigh it against their modifiable exposome, they may be more likely to place it in the proper context.

While many of the issues and concerns mentioned above are addressed in the behavioral sciences and risk science, the general public is not so concerned with decision-making processes, biases, and probabilities. We know that most people are not intuitive risk scientists, they are looking for an easy way to frame the complexity with which they are bombarded. The exposome provides this framework and can introduce biological concepts in a context that makes it easier to remember over time. One does not need a college course in biology to understand the key components. When it comes to exposures that affect our health, the exposome stands out as a workable and teachable model that can be used to improve health.

6.9 OBSTACLES AND OPPORTUNITIES

The transfer of knowledge about the exposome should not be difficult from an educational standpoint. The primary challenge for the field is the general acceptance of the concept itself. Members of the general public are inclined to embrace a holistic view of their health, and the exposome concept fits within this view. However, adoption of the exposome within the halls of academia requires those in the field of environmental health sciences to see the value of a unifying framework that expands beyond their traditional intellectual boundaries. Within the field of environmental health sciences, many of the concepts of the exposome are already being addressed even if the term is not being used, or if the vision is not as broad. Incorporating the exposome, especially the big data-generating approaches, into courses within the environmental health sciences is necessary. The explosion of complex

Learning Objectives

Chapter 1. To develop an explanation of the exposome concept in the reader's own words that can be used to explain the exposome to scientific and non-scientific colleagues.

Chapter 2. To be able to summarize the utility of genetic studies in the study of human disease, as well as the limitations.

Chapter 3. To define the major -omic disciplines, including the major techniques used in each.

Chapter 4. To be able to explain how the various subdisciplines in environmental health sciences can contribute to acquisition and dissemination of exposome-related research.

Chapter 5. To provide specific examples and strategies of how tools from systems and computational biology can be used in the analysis and integration of exposome-related data. In addition, the student should be able to demonstrate how maps and spatial statistics are used in the organization of exposome data.

Chapter 6. To be able to incorporate exposome-related concepts into discussions with colleagues and members of the general public as a way to promote the importance of the environment in our health.

Chapter 7. To be able to design and defend a framework for an exposome initiative (e.g. HEP) and to provide specific examples of potential impediments or limitations of such an undertaking.

Figure 6.3 Learning objectives for The Exposome: A Primer. From an educational perspective, it is imperative to have learning objectives that outline the key skills and knowledge that are to be transferred to the student. The list here corresponds to this book, but depending on the content of a course (Figure 6.2) the objectives may vary.

data sets is not going to slow down anytime soon. In fact, trainees that ignore this unavoidable truism will be at a significant disadvantage. Integration of the exposome concept into medical education may be a different matter. The medical profession is very thoughtful and careful about alterations in their information-dense curriculum. It may be best to start with continuing education (CE). Developing CE courses that address certain aspects of the exposome could become a useful tool in medical education. However, improving environmental health science education among medical professionals is an important and attainable goal and those with expertise in environmental health sciences should continue to explore ways of doing so. The field will need to prepare and deliver the appropriate type of materials for education and training for each of the aforementioned areas. The more resources that can be provided to the professor, scientist, or teacher willing to present this material, the more likely the effort will be successful. Ideally, courses that incorporate the exposome theme will become common in academia. With concerted efforts from scientists and public health professionals, the exposome can become a useful vehicle for teaching the importance of the environment in our health to those in other scientific disciplines and to the general public.

6.10 DISCUSSION QUESTIONS

Figure 6.3 shows learning objectives for the exposome. How have you fared? Do these objectives capture the essence of the material? What would you change?

Create a new version of the modifiable exposome with community on the left side of the x-axis and society on the right side. In essence, remove the role of the individual. Conversely, one can create a new version that plots individual on the left side of the x-axis and community or family on the right side. Discuss the differences and similarities among the different exposome plots.

Outside of class explain the concept of the exposome to a scientific colleague and a nonscientific colleague. What was the response?

FURTHER READING

Ahmed SM, Palermo AG. Community engagement in research: frameworks for education and peer review. Am. J. Public Health August 2010;100(8):1380−7. doi:10.2105/AJPH.2009.178137.

Burstein JM, Levy BS. The teaching of occupational health in US medical schools: little improvement in 9 years. Am. J. Public Health May 1994;84(5):846−9. doi:10.2105/AJPH.84.5.846 This paper describes the reduction of occupational health programs in the US. While the article is 20 years old, it is safe to say that the situation has only gotten worse since then.

CHAPTER 7

Staging the Exposome: A Vision for International Collaboration

Over the past years, I have often been asked "How does one go about measuring the exposome?" or "What would a Human Exposome Project look like?" I typically defer answering this type of question because of the sheer complexity of the problem and an inability to explain a massive project in just a few sentences. It is good that the questions are being asked. If people are interested in how such a project would be pursued, it is likely that they view it as an area worthy of inquiry. This chapter is my first attempt to answer these questions.

7.1 MEASURING THE EXPOSOME

In a manner analogous to the Human Genome Project, a well-coordinated international collaboration will need to be established to begin the process of measuring the exposome. This chapter outlines a possible strategy for moving a Human Exposome Project forward, including the importance of data sharing and international collaboration. I am a fan of the word collaboration, but not so much of the word consensus. Indeed, I shy away from the use of the word. Consensus means broad unanimity, as in unanimous, as in everyone agrees. If everyone agrees on how to approach the exposome, then we are doing it incorrectly. There will be and should be significant disagreement and debate about how to do this. Consensus is also often used in discussion of professional opinion. The exposome should not dabble in opinion. The exposome must be data driven and verifiable. Consensus is also often used to suggest that research is no longer needed, that the scientific overlords have spoken. This is a very dangerous position in which to be. As scientists we should welcome thoughtful criticism and a continued degree of skepticism. It is not easy to develop cogent arguments and valid scientific methods. The continued consideration of opposing views and approaches helps sharpen scientific thinking. The key is to create an inclusive process that accepts the input of representatives from multiple organizations, but ultimately comes up with a process that is acceptable to at least a majority of participants and is scientifically sound to an impartial observer. Objections should be noted and considered throughout the process. Alternative approaches can be pursued and the results compared to the majority view.

A single author proposing a plan for how to pursue the exposome is the antithesis of the aforementioned plan. However, for the purpose of discussion and in the hopes of spurring lively debate, this is precisely what I will do here. First off, exposome research should involve a series of coordinated and parallel projects run from different sites throughout the world. It should not be one single massive project. There are many steps that can be performed at the same time. As shown in Figure 7.1, we are currently at the stage of developing and testing enabling technologies.

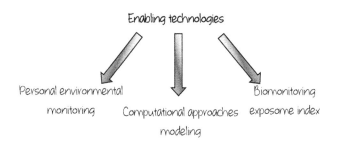

Figure 7.1 Enabling technologies. Exposome research requires ongoing development and refinement of methods and approaches to improve our ability to identify exposures. Continued development of personal environmental monitoring, biomonitoring, and computional platforms will help provide the field with the necessary tools to move forward.

The current exposome-related projects are building the infrastructure, testing out new methodologies, and determining if the improved resolution of exposures is adding value to the study of human health and disease. These efforts are occurring in parallel, but there has been little interaction among these efforts.

Exposome-related research and a Human Exposome Project are two distinct activities. The former is underway while the latter is a goal at this time. The work currently being conducted is providing the foundation for an eventual Human Exposome Project. Thus, one could divide exposome research into two divisions. The first is the work that is conducted up to the point that a true Human Exposome Project is

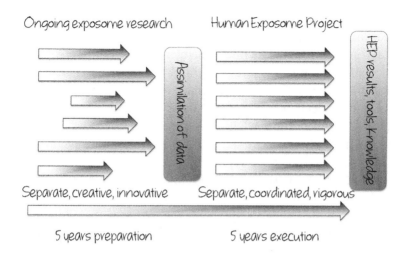

Figure 7.2 Proposed staging of a Human Exposome Project. Pursuit of the exposome will require a highly coordinated effort with international partners. Here is a possible scenario and timeline of the steps that could be followed in the development and pursuit of such an undertaking.

initiated. The second is the Human Exposome Project itself. As depicted in Figure 7.2, there are numerous exposome-related activities underway, some of which are described below. The idea would be to use the data and tools being generated by these ongoing projects to design and conduct a full-scale Human Exposome Project.

Increased communication among the various investigators should occur, but during this buildup phase nothing should be dictated. The various groups should be free to conduct their work as they see fit. In essence, all of the ongoing research will serve a series of pilot studies to determine the best tools and approaches to be used in more comprehensive projects in the future. However, these groups should start meeting on a regular basis to exchange data and insight. The stronger the collaborative relationships are over the coming years, the higher likelihood that the groups will be willing to enter into a multisite project to conduct the large-scale studies that will be necessary.

7.2 WHAT IS IN US? WHAT IS NEAR US? DOES IT MATTER?

At this point it will be helpful to review what it is we are trying to get out of the exposome. What should be the focus? Obviously, the exposome is going to be heavily focused on the chemicals that reside or did reside in our bodies. We want to know what chemicals should be in us and what chemicals should not be in us. We want to know how the chemicals alter our biology. We want to know how exposure to a class of chemicals impacts our endocrine system, modifies our DNA, or alters synaptic signaling. We are developing and testing sophisticated methods for measuring the biological soup, and these approaches will greatly enhance our understanding of how these chemicals are affecting our biology. We want to know what is near us. In a systematic and parallel fashion, we must also measure the complex exposures outside of the body (what Dr. Lioy and the NRC report referred to as the eco-exposome, but that I consider to be part of the exposome). Where are these chemicals coming from? Does living ten miles from a factory actually increase the deposition of emitted chemicals to a person? What if they drive past the factory everyday? Does it matter if the exposure occurred in combination with another pollutant or nutrient? Without concomitant knowledge of the source of the chemicals, knowledge of which chemicals are in our bodies and what biological effects they are exerting is limited in application. Figure 7.3 provides a sketch of the complex layers of exposome-related data. We will be generating data across various domains, but those data must be used to inform the other domains. We will be dealing with data from molecular, cellular, animal, human, population, and environmental sources. We must work on integrating these complementary findings in ways that improve our overall understanding of how environmental factors alter our bodies. Investigators must be willing to work among the various domains. This will require the use of sophisticated computational approaches as addressed in Chapter 5.

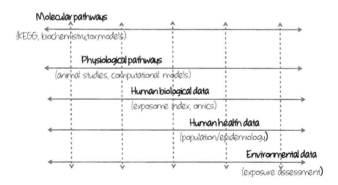

Figure 7.3 The multi-tiered organization of the exposome. The exposome will require input at multiple levels including molecular pathways, physiological systems, organismal, population, and external environment. Data from each level can be used to inform the other levels. Organization of multiple computational models will be required to capture the expansive nature of the exposome.

Does it matter? The question of whether or not the chemicals that are present in our environment and in our bodies actually harm our health is what the exposome is all about. As noted in Dr. Wild's original paper, epidemiologists need a higher quality of data to make such determinations. The exposome should be able to provide this higher fidelity data, but the disease association studies will still need to be conducted. It may be that we need a few more years of technological development before we have the type of data to achieve Dr. Wild's vision.

7.3 A BASIS IN BIOLOGY

The exposome should be viewed as a biological challenge. We want to know how the environment influences our biology. We must build upon our current understanding of biological systems. Biochemical pathways, physiological systems, and mechanisms of toxic action must be the foundation of the exposome. There must be biological plausibility at every step (although a level of biological uncertainty could still be factored in). The molecular pathway model is an excellent starting point, in that much of the key information is already available. Basing a model upon the known biological pathways and networks grounds the model in an already characterized system (note how this parallels the PoTs approach described for the toxome). For example, the KEGG pathways could be used as a starting point. It could use biological pathways that have already been identified to be perturbed by environmental insults, such as key endocrine pathways that can estimate biological outputs from alterations in estrogen and testosterone receptor binding and activation. Structure activity/toxicophores could be built that would allow for multiple compounds to be screened *in silico*. Computational models from multiple pathways can be assembled together into more sophisticated models, as discussed in Chapter 5. While the Tox21 initiative is built upon molecular and cellular pathways, the motivation for it comes from more of a regulatory perspective. While the exposome could have a major impact on regulatory decision-making, such issues should not be the basis for conducting the research. If the exposome is perceived as a mechanism to increase regulation, it will face strong resistance from various groups that must be involved in the long-term solutions. We should be evaluating the lifespan of the chemicals that are doing us harm. What are the needs and uses of the goods that contain or require the chemicals?

Are there alternate manufacturing processes that can reduce emissions and exposures? If we want to optimize health and minimize disease and suffering, we must involve industry and manufacturing partners in the solutions. While many questioned the approach and cost of the Human Genome Project, few questioned its scientific value and it was supported by both sides of the political spectrum. The exposome should be viewed in a similar manner. The exposome will be providing insight into the nongenetic factors involved in health and disease. We need this information. The focus should be on the acquisition of knowledge.

7.4 IT HAS ALREADY STARTED

When I prepared the syllabus for our first exposome-based course, there were no major exposome-based research projects being conducted. Over the past year, several major projects have been funded. It is clear that the European Union is taking the lead. A brief overview of some of these projects is provided here, and links are provided at the end of this chapter.

HELIX: Building the Early-Life Exposome is a Barcelona-based research project that is focusing on the impact of environmental exposures on early life stages. The project has several objectives. One is to measure chemical and physical hazards in environmental matrices (food, water, air, consumer products, built environment) during pre- and postnatal periods. The project will also assess multiple exposure patterns, individual variability, and quantify uncertainties. The project will also examine molecular and biological profiles to estimate the relationship between exposures and outcomes in childhood, and estimate the burden of disease due to environmental factors. The group hopes that this information can be used to inform European policy regarding the various exogenous factors.

The Exposomics Project based in the UK is a multi-institution consortium that will emphasize improved portable technology for monitoring personal exposures. The Exposomics Project is taking advantage of many of the recent personal environmental monitoring advances. Using handheld smartphone devices with global positioning technology and environmental sensors, the investigators will track subjects' movements and exposures during their day-to-day activity. The effort to develop and validate the portable environmental monitoring systems could help make such technology more broadly available. The investigators will also be looking for various signatures of exposures by analyzing DNA, RNA, protein, metabolites, and chemicals in the blood. There are also personal portable devices that measure carbon dioxide, temperature, humidity, particulate matter, nitrates, noise, and radiation. Before long it will be possible to incorporate all of these and more into smartphone-based systems that also track movements using GPS technology, and the Exposomics Project is driving these types of technology forward. There is also technology for caloric intake and expenditure and prompting the user to report of their subjective feelings of well-being. Encouraging further development and miniaturization of such technologies will be a boon for exposome research.

Another recently funded European center is the HEALS project, which stands for *Health and Environment-Wide Association Studies based on Large Population Survey*. HEALS is working to refine analytical and computational tools to perform EWAS-type studies. HEALS will use preexisting population data along with newly collected environmental and health data from 18 EU member states. The consortium will utilize geospatial, socioeconomic, and health data along with -omic, remote sensing, microsensors, and systems biology approaches to provide exposome-level data in these large population studies.

The *HERCULES: Health and Exposome Research Center at Emory* was recently funded by the US National Institute of Environmental Health Sciences. Unlike the HELIX and Exposomics projects, HERCULES is a core center mechanism that provides infrastructure needed for investigators working in the area of environmental health science. HERCULES is focused on building infrastructure for conducting future research on the exposome. Service cores provide expertise in systems biology, bioinformatics, analytical chemistry, and metabolomics. Pilot grants will be awarded each year to encourage investigators to incorporate exposome-related techniques and concepts into their research. There is also an emphasis on getting investigators to deposit existing and ongoing data into databases that can be mined.

Each of these ongoing studies is providing new information that will contribute to our understanding of the exposome. Ultimately, we would want to use the data we have to determine what the ideal experiment approach would be. We need to know what we are missing. We need to determine the sensitivity of our measures and the ability to determine cause and effect. We want to be able to develop a series of studies that provides the appropriate power, comprehensive and repeated analysis, robust biological and physiological data, that can be replicated in multiple sites across the world.

7.5 SO WHAT ARE THE EXPOSOME DELIVERABLES?

There are several potential outcomes of exposome research. The first major deliverable will be improved methods for measuring and cataloging the chemicals in our body. This will likely be the result of improved mass spectrometric technologies and other techniques that provide exquisite separation and identification in biological samples. The next major deliverable will be validated measures of the impact of exposures on our biology, that is, the cumulative biological responses in the revised definition. This will likely come from a combination of approaches including some of the newer methods to assess DNA modifications at a genome-wide scale. Enhanced exposure assessment from improved technology and data will benefit from being collected at the same time that data from the internal measures are being collected. In essence, by being yoked to the biological measures, the data from the exposure assessment will yield a new dimension of insight. Another key outcome will be a mechanistically based computational platform(s) that enhances our understanding of how suspected chemicals (thousands) contribute to disease. Much of this may be built upon systems-biology-based models of human diseases being generated by other fields. The key will be to develop the interface that allows the impact of the environmental chemicals to be assessed by the model. One of the most important deliverables will be a suite of measures (an exposome test or index, or a battery of tests) that can be used in population studies to examine the relationship between complex exposures and human disease that could be used in a clinical setting to improve health care.

7.6 GENE × ENVIRONMENT × GRANTS

The European Union is making a major investment in exposome-related research, but a similar level of effort is not seen in the US. In the introduction, I addressed the disadvantaged allocation of NIEHS in the US. I am not advocating equal distribution among all branches of NIH but I certainly will argue that research on the role of the environment in health and disease must be expanded. Within the US, NIEHS should be receiving 1/20th of the NIH budget *at a minimum*. The distribution of research funds is highly influenced by lobbying groups. It is difficult to criticize an organization that advocates for increased funding for a disease that has impacted the lives of their members, but this is one of the problems with research surrounding environmental influences of health. It is not a single disease or diagnosis. The impact of the environment is strewn across multiple (nearly all) disease states. Allocating only 5% of NIH research to environmental causation of disease is underwhelming to say the least and such restructuring of NIH dollars is fanciful at best. One may argue that other institutes do study environmental factors, but not to any significant degree. Perhaps the approach is not to increase funding of the environmental health-based funding institutes, but rather increasing allocation to environmental health or exposome-based research at other research funding institutes. However, it is very difficult to argue for a research institute to expand their environmental health or exposome-based portfolios when the tools are not yet available. For long-term impact providing sensitive and validated tools and concepts that allow investigators outside the field to integrate the environment into their research should be a top priority. While allocations across institutes may not change, if institutes recognize the potential utility of the exposome for their discipline and there is a framework for conducting such research it is not ridiculous to think that other research institutes could start encouraging work that incorporated the exposome.

Within the exposome framework, many research institutes are studying environmental factors. Studies that involve dietary influences, behavioral modifications, and social determinants of health, factors outside the traditional scope of environmental health science research institutes, are underway at other institutes. The more politically palatable solution may be to integrate more environmental health-related research into those research institutes and funding agencies that study diseases that are suspected of having a strong environmental component. One could argue that environmental health-based research institutes are a bad idea because it gives other groups an easy way out of addressing environmental factors. However, if one were to close such an institute and redistribute its funds across the others to conduct environmental health-related research, only those exposures that have extremely strong relationships would be studied. The subtle influences on health would be completely missed.

I wrote this text from an academic perspective. The goal is to understand how our surroundings influence our health from a biological and scientific standpoint. But we do live in a practical world with a need for practical decisions. Regulatory agencies in the US, Europe, and other parts of the world must make decisions about environmental factors. How can these groups use the exposome? One of the challenges of regulatory agencies is that they are in that tricky business of prediction. They are trying to predict which factors are most likely to be contributing to human disease and then predicting which interventions will best mitigate these risks. Many of these agencies are looking to exposome-type projects to guide them in their predictions but, as was discussed in Chapter 5, this is a bit risky. That said, many of these agencies have collected massive amounts of data that are very relevant to the exposome, and they are proceeding with projects to develop predictive algorithms. This may be a bit premature. We do not know if the data we are collecting is scalable. Can our current models adapt the complex data being generated? The science described in this text should help inform these agencies, but until the approaches are empirically tested it may be unwise to integrate them into regulatory decisions.

7.7 AN EXPOSOME INDEX

A major goal for the exposome should be for its concept and information to be integrated into the care of the patient. This may be one of the harder goals to achieve, but it must remain a goal. As noted previously, it would be preferable to focus assessing current abilities to respond to insults and not on predicting future outcomes. In essence, how robust is an individual based on the current assessment of his or her exposome? Twenty years ago, the term allostasis was introduced by McEwen and Stellar as a way to describe the wear and tear that occurs in the body in response to repeated stress. Allostatic load has been used in some fields, but it has not gained widespread use. Some of the concern is that the measurements are somewhat crude, though readily measurable (systolic and diastolic blood pressure, waist hip ratio, serum HLC and total cholesterol, plasma levels of glycosylated hemoglobin, serum dihydroepiandrosterone sulfate (DHEA-S), overnight urinary excretion levels of cortisol, norepinephrine, and epinephrine). While these measures can provide an assessment of the hypothalamic–pituitary axis and responses to stress, it does not get to the issue of "cumulative … associated biological response" in our definition of the exposome. For this we must measure some biological changes that persist over time, such as epigenetic changes, DNA adduct formation, telomere length, and other aspects of cellular damage. Approaches that measure the body burden of environmental chemicals can provide an index of exposure, and when compared to the relative "associated biological response" one can assess the relative ability of an individual to respond to subsequent insults. Thus, for the exposome we need to ascertain the most appropriate battery of tests to measure this cumulative biological response that would be analogous to the allostatic load. Combined with a comprehensive assessment of exposures, such a measure could become part of an exposome index that could ultimately be used in a clinical setting. For the purpose of discussion, Figure 7.4 shows a cocktail napkin sketch of a hypothetical exposome index.

Figure 7.4 The back of the napkin exposome index. As a means of stimulating discussion, a list of possible components to an exposome index is shown. The idea is that each of these components could be measured in an individual at a specific point in time. Future studies would need to empirically determine whether or not these values correspond to changes in health status, which would allow refinement and retesting.

This is just an initial idea of a suite of tests that could be run to assess exposome-related exposures and biological effects. Our DNA stays with us for life (although the cells harboring it may turnover, which makes the source of this DNA critical). While DNA repair mechanisms are continually working to maintain the integrity of our genetic code, there are lasting alterations to our DNA in the form of epigenetic modifications, adducts, or outright damage (deletions, cross-linking, etc.). Telomere integrity is considered to be one way to assess the health of our DNA. Thus, measurement of several markers of DNA integrity provides a measure of our lifelong genetic insults and our ability to respond to said insults. What is measured on a particular day is the summation of these injuries, repairs, and regulatory responses. Our exposure to chemicals, including the absorption, metabolism, distribution, and excretion of them, is one of the most important aspects of the exposome. When we think about how to reduce our exposures to certain chemicals, we need to know which chemicals end up in our bodies relative to the presumed exposures (from use, or air, soil, or food sampling data). The suite of chemicals in our bodies is the result of not only our own metabolic processes, but that of our microbiome. Is it necessary, though, to measure every possible chemical? Given that there are over 100,000 manmade chemicals in our environment this would seem to be an impossible and unnecessary task. It would be better to identify a set of chemicals or chemical classes that are representative of our complex exposures. Development of a high-throughput platform to measure major classes of chemicals would be an important step in defining the exposome.

While an exposome index addresses many of the body's responses to exposures, it does not capture the source of many of the external influences. For example, physical activity can be addressed by personal accelerometers in smartphones or similar devices. However, gleaning information on social determinants of health will be more challenging. The impact of stressful life experiences can be captured as part of the assessment of epigenetics, but determining the occurrence of such a life stress is more challenging and will require input from the subject. One of the interesting aspects of an exposome index is the ability to test its validity with tools that are already at our disposal. Prospective epidemiological studies could be designed to test the predictive validity of certain components of the index. It would be best to focus on health outcomes that could be measured over the course of a few years. Development of an initial exposome index could be a useful topic for an initial exposome meeting.

7.8 A HUMAN EXPOSOME PROJECT?

Is it too early to start discussing a Human Exposome Project? It is not clear if we are ready for our Santa Fe meeting to launch such an initiative. More proof-of-principle data must be collected, but it is time to start having the discussion. It is time to provide the complementary environmental analogue to the Human Genome Project. Whether it achieves the level of being a biological index of nurture is not clear, but there is no doubt we need to focus more resources toward our environment. In 5 years, initiation of a Human Exposome Project is feasible, but it will take a considerable amount of careful thought and planning.

As shown in Figure 7.5 (and Figure 7.2), planning the execution of a Human Exposome Project is a long-term enterprise. One of most exciting aspects of the outline shown is that so many of the activities are already occurring. The enabling technologies are under development. The exposome is being discussed at annual scientific conferences across multiple disciplines. There is more frequent exchange among the research groups across the world. The interest in exposome-related research is growing, resources are being allocated, and energy is being devoted to the idea. Momentum for the exposome is building. The

Human Exposome Project (HEP)

2014-convene initial planning meeting with international partners from academia, industry, government, and private sectors, form executive committee (HEPEC)

2015-cont. development of enabling technologies, HEPEC meeting, garner resources

2016-cont. development of enabling technologies, HEPEC meeting, garner resources

2017-cont. development of enabling technologies, pilot exposome index components, HEPEC meeting, garner resources

2018-finalize design of HEP project, HEPEC meeting, garner resources

2019- initiation of large, five-year, multi-site prospective cohort study that utilizes the best of the technologies developed and tested to date

2020-onward-frequent data and progress updates, HEPEC meetings, dissemination of findings

Figure 7.5 Ten-year plan for the Human Exposome Project. There are several exposome-related projects underway. Over the next few years, these unrelated projects will identify the best technologies and approaches. However, a Human Exposome Project will require coordination among the various groups. The goal is for the groups to pursue the second round of studies in a manner that allows combining of all of the data into a single exposome model or database. This would increase the statistical power, provide geographic and genetic diversity, and would leverage the resources of all of the various research centers.

goal should be to harness the results of these ongoing activities, provide a clear focus for ongoing research, and then work to direct future initiatives that allow us to pursue the most exciting and informative science.

The recent awarding of the international centers mentioned above would seem to provide a superb foundation for planning a more expansive initiative. The formation of the Human Exposome Project Executive Committee (HEPEC) will be vital to the success of the initiative. It should be composed of energetic and creative individuals willing to collaborate. The key drivers of current exposome activities should be represented, as should the representatives of the funding organizations. There will be a need for frequent interactions in the form of teleconferences, list services, and web-based updates. The convening of an initial kick-off meeting will be critical. Getting the right people together to establish the vision and outline a plan is essential.

The HEPEC will need to address issues surrounding data sharing and authorship policies. Data hosting sites and systems will need to be identified (and funded). Government science agencies will likely provide the majority of funding, but it will be very important to establish partnerships with industry to assist with the technological demands, as well as working with private partners to raise funds for research and planning. Private support could help accelerate some of the initial brainstorming and planning.

A variety of studies and approaches will proceed over the next few years. The goal is to identify the best approaches being used by the various research projects and integrate them into the second phase of coordinated projects. For example, if a group has developed a novel personal environmental sensor, it would be ideal for all groups to employ such devices in the coordinated phase. Common approaches for identifying DNA adducts, performing metabolomics, or measurement of specific chemicals could be adopted. In the illustration six arrows or research centers are depicted, but this could be higher or lower. A core-based approach may be warranted in which many of these tests can be run at two separate sites for all of the human exposome project endeavors, with each site providing a particular expertise. In a similar fashion, investigators may want to use traditional survey instruments to determine how well these parallel the newer exposome-based approaches, but each site should be using the same instrument.

To take such an approach, the various funding agencies would need to agree to the collaborative approach, which has precedence. It may be more difficult getting the various scientists and teams to agree on common approaches. As noted before, there will be significant agreement on what is the best way to address certain problems, but it will be critical to focus on the core problem of better understanding the environmental influences on our health. A large, well-designed, and rigorously executed study could have a dramatic influence on further studies in science. Thus, even after the end of the proposed 10-year plan there will be more to do. Ideally, the research conducted over those 10 years makes society better appreciate how important the environment is to their health and helps put the environment in a more prominent place on the research agenda. In 10 years, we could have a new suite of tools and technologies that can be employed in other fields that have traditionally ignored the environment, have the preliminary data for examining environmental factors for diseases that have not considered environmental factors, and have permanently added the word exposome into the biomedical vernacular, and its concepts into health care models and into the minds of the general public.

7.9 OBSTACLES AND OPPORTUNITIES

Alas, the exposome faces many obstacles. As outlined in the above chapters, the exposome is complicated and will not readily reveal its secrets. However, we are making progress. The exposome provides numerous opportunities and challenges for investigators interested in the influence of our environment in health and disease. Even if naysayers wanted to dispatch with the entire unifying notion of the exposome and jettison the word from the field, it would be difficult to do so. It is clear that pursuing a better understanding of the complex and lifelong exposures that impact our health is a worthy goal and the exposome encapsulates that goal. Environmental health sciences and exposome-aficionados must be willing to address the issues surrounding the acquisition, analysis, and interpretation of big data, including the need to cultivate investigators who possess the necessary skills to perform the required studies. Without an aggressive campaign to recruit and develop these types of thinkers, environmental health sciences will continue to play the role of follower within the biomedical community instead of the role of leader. Creative scientists with training in bioinformatics, systems and computational biology, must be lured into the field. We must start developing the computational platforms that will help us start sorting and organizing the bricks that are already piling up, but more importantly, to organize the carefully constructed bricks yet to come. We simply cannot proceed until we organize the brickyard.

Does the exposome represent a sufficient challenge to entice the next generation of scientists? From my perspective, the answer is a definitive yes. Indeed, the vision for the exposome and its associated complexity can attract the ambitious young investigators. While the limitations of the gene-centric view of health and disease may have been intuitively obvious for those in environmental health science, they are becoming demonstrably so to the rest of the scientific community. The exposome provides the framework for the next major advance in the study of human health and disease. The exposome will also need specialists in the core subdisciplines, but such investigators would be wise to expand their scientific repertoire to include computational approaches. Perhaps some of the young and energetic readers of this book will recognize the exciting opportunities provided by the exposome, start contemplating specific challenges, and eventually help define and refine this burgeoning concept.

7.10 DISCUSSION QUESTIONS

Refer to the napkin figure (Figure 7.4). What would you add to or delete from the proposed exposome index?

Does such an exposome index provide a measure of nurture?

Does it make sense to work toward a Human Exposome Project or is it better to continue current lines of research and develop the human exposome after more data have accumulated and technology has advanced?

FURTHER READING

Collins FS, Morgan M, Patrinos A. The human genome project: lessons from large-scale biology. Science 2003;300:286–90.

Kanehisa M, Goto S, Sato Y, Furumichi M, Tanabe M. KEGG for integration and interpretation of large-scale molecular datasets. Nucleic Acids Res. 2012;40:D109–14.

Wild PC. The exposome: from concept to utility. Int. J. Epidemiol. 2012;41:24–31.

ADDITIONAL RESOURCES

Human Exposome Project—Humanexposomeproject.com

Exposome Alliance Project—http://exposomealliance.org

http://www.exposomicsproject.eu

EnviroGenomarkers: Genomics biomarkers of environmental health, a European Union-funded center—http://www.envirogenomarkers.net

Developmental Neurotoxicity Assessment of Mixtures in Children, a European Union-funded center—http://www.denamic-project.eu

HELIX: Building the early life exposome—http://www.projecthelix.eu

Health and Environment-wide Associations based on Large Population Surveys—http://www.enve-lab.eu/index.php/enve-lab-leads-heals/

Public Health Exposome—http://communitymappingforhealthequity.org/public-health-expo-some-data/

NAS website on exposome meeting—http://nas-sites.org/emergingscience/meetings/exposome/

Wikipedia—http://en.wikipedia.org/wiki/Exposome

US Environmental Protection Agency—www.epa.gov

European Environmental Agency—http://www.eea.europa.eu

Ministry of Environmental Protection, The People's Republic of China—http://english.mep.gov.cn

US National Institute of Environmental Health Sciences—www.niehs.nih.gov

US Environmental Protection Agency Computational Toxicology/Tox21 initiative—http://epa.gov/ncct/Tox21/

Printed and bound by CPI Group (UK) Ltd, Croydon, CR0 4YY

03/10/2024

01040427-0013